John Gregory

A Comparative View of the State and Faculties of Man with Those of the Animal World

Eighth Edition

John Gregory

A Comparative View of the State and Faculties of Man with Those of the Animal World
Eighth Edition

ISBN/EAN: 9783337805630

Printed in Europe, USA, Canada, Australia, Japan

Cover: Foto ©berggeist007 / pixelio.de

More available books at **www.hansebooks.com**

A

COMPARATIVE VIEW

OF THE

State and Faculties of MAN

WITH THOSE OF THE

ANIMAL WORLD.

By JOHN GREGORY, M.D. F.R.S.
Profeſſor of Medicine in the Univerſity of Edinburgh,
and Firſt Phyſician to his Majeſty in Scotland.

THE EIGHTH EDITION.

DUBLIN:

Printed for W. SLEATER, D. CHAMBERLAINE,
J. POTTS, J. WILLIAMS, and W. COLLES.
M.DCC.LXXVIII.

CONTENTS.

PREFACE.

APOLOGY for the work and for the Title of it, page i.—The connection of the Subjects treated in it, to be explained from the state of manners in the different periods of society. Man when a savage, iii. Man when in the full vigour of his faculties, iv. Exemplified from the state of manners described by Ossian, v. Progress and corruption of Society, ix. Man when enervated by wealth and luxury, xi. Though large societies cannot, that possibly individuals may unite the advantages which arise in the various stages of Human manners, xiv. That with views to promote this end among a few friends the work was originally composed, xv.

SECTION I.

DIFFERENT views of human nature, page 1. The difficulty of prosecuting enquiries into the constitution of the Human Mind, 4.—The philosophy of the Human Body and Human Mind must be united, to be prosecuted with success, 5. Comparative views of the state and manner of life of Men and Animals, not suffi-

CONTENTS.

ciently attended to, 7. Distance between the Faculties of Man and Animals, 9. — Cause of the want of Language in Animals, *ib.* — Pleasures peculiar to the Human Species, 10. Advantages enjoyed by lower Animals, 12. The separate provinces of Instinct and Reason, 13. Importance of ascertaining the natural Instincts of Man; and for this end to enquire into the analogous Instincts of other Animals, 14. The breed of Animals may be improved or debased by art: May not attention do the same among Men? Family characters and Constitutions distinguishable, 17. Greater Mortality among Children, than the young of other Animals, 20. This evil neglected because it is common, 21. Occasioned by forsaking the dictates of Instinct and Nature, 23. The advantages peculiar to Children in point of constitution, *ib.* Evil of giving Children physic at their birth, 25. Of delaying giving them suck, 26. Of Mothers not nursing their own Infants, 28. This subject of Nursing considered in many points of view, down to 36. Evil of confining Children, 37. Cloaths, air, exercise, considered, down to 42. Evil of suffering Children to cry, and of quieting them by rocking, 43. Rocking in every case improper, 44. Evil of over-feeding, *ib.* Evil of keeping Children too warm, 46. Bad effects of hot Air and Regimen on lying-in women, 47. The activity of Children, 51. To be indulged, 52. The country the proper place for the education of Children, 54. Strength and hardiness of the Human constitution, 56. Our manners exposed to the diseases of our climate, 58. Teething chiefly dangerous from the errors of modern education, *ib.* Evil of too early study, 60. The order of nature, in developing the Human Faculties, to be followed

CONTENTS.

ed in education, 61. Attention and abilities required to educate children, 64. Conclusion, that many of the evils we suffer arise from neglecting Nature; and in particular, that the chief miseries usually attendant on old age, proceed from an unnatural manner of life, 65.

SECTION II.

The faculties which chiefly render Man superior to other Animals, to be considered, page 69. Reason not a blessing unless properly employed, 70. Of Genius and a superior Understanding, *ib.* Seldom employed in promoting the useful Arts, 71. Wasted on idle Theories, *ib.* This illustrated from the state of Medicine, 75. Also from the state of Agriculture, 76. How Science to be promoted, 77. Bad effects of the passion for universal knowledge, 81. Superior talents as respecting the happiness of the possessor, 82. Unfavourable to the social affections, 83. Other effects of that abstraction from company, which they often occasion, 88. Bad Health, 89. Scepticism, 90. Solitude, 91. Social principle, 93. Not naturally powerful in our climate, 94. Hence, to render love powerful and permanent here, it must be connected with friendship, 97. The situation of women considered, down to 100. The circumstances which render life agreeable, much neglected, 101. The association of different ages and sexes, promote the common happiness of the whole, 104.

SEC-

CONTENTS.

SECTION III.

Of Taste, 109. Must be cultivated by philosophy before it can afford much enjoyment, 111. Just criticism, much neglected in Britain, 112. Bad effects of such neglect on Painting, 113. Bad effects of it on Music, 114. Music particularly considered, 117. Object of Music, 118. Music among uncultivated nations, 119. Progress of Music illustrated from its History among the Greeks, 120. The effect of Music on eloquence, 124. Of national Music, 128. Little influence of modern Music, 133. Simplicity of melody necessary to affect the Heart by Music, ib. Different circumstances which render Music expressive of the passions, 137. Hence the influence of the ancient Music accounted for, 139. Of Harmony in Music, 141. The present stile of Music, 146. Music considered as an amusement only, 148. National Music influenced by the poetry of the country, 152. Circumstances to be attended to, in the composition and performance of Music, 155.

SECTION IV.

Works of Genius which are objects of Taste to be considered, page 167. The application of the powers of the Understanding, to the reducing of Taste to strict rules, difficult, and often pernicious, 169. Taste different in different nations, 171. Correct Taste not favourable to Genius, 172. Evil of over-refinement, 173. Character of the French and English Taste, 175. Difference between probability addressed to the Understanding, and adopted by the Imagination,

CONTENTS.

gination, 177. Effects of ancient Romances and modern novels, 180. History only interesting when it engages us for some public cause, or distinguished character, 182. Its best end defeated, if the Historian prevents this effect, *ib.* Amusement but a secondary end of History, it should warm the Heart to liberty and virtue, 183. Imagery, 186. Wit and Humour, 188. Mode of contemplating the objects of Taste with most advantage, 189. Pleasures of Sympathy, 193. Object of cultivating Taste, 194.

SECTION V.

The advantages of Religion, page 197. Corruption of it, 198. Atheism and Superstition compared, 200. Sense of right and wrong, independent of Religion, 201. Infidelity owing to insensibility of Heart, 203. Religion not founded on weakness of Mind, *ib.* Effects of Religion, Scepticism, and Infidelity, 204. Comforts of Religion, 209. Cause of zeal to propagate Infidelity, 210. Such zeal inexcusable, 214. Religion considered as a science, *ib.* Religion considered as a rule of Life and Manners, 222. How it is to be applied to cure the diseases of the Mind, 224. Public addresses, 227. Religion considered as exciting Devotion, 232. Advantages of Devotion, 234. Conclusion.

PREFACE.

BY an advertifement prefixed to the firft edition of this book, the public was informed that it confifted of fome difcourfes originally read in a private literary fociety, without the moft diftant view to their publication. The loofe and carelefs manner in which they are written, is too ftrong an internal evidence that they never were intended for the public infpection. But, for what purpofe they were originally compofed, and how they came into the world, are queftions which a reader will never afk: he has an undoubted right to cenfure them with all the feverity which their faults deferve, and to cenfure likewife the author of them, unlefs he could pretend they were publifhed without his knowledge. The unexpected favor he has met with from the public has encouraged him to correct and enlarge this edition; but when he attempted to treat

PREFACE.

treat his subject with that fullness and accuracy which its importance required, he found it run into so great an extent, that he was obliged to abandon it, being necessarily engaged in business and studies of a very different nature. He would gladly have suppressed some sentiments carelesly thrown out in the confidence of private friendship, which may be liable to misconstruction; but he was afraid that, by too anxious an attention to guard against every objection, he should deprive the book of that appearance of ease and freedom in which its only merit consisted. When we unbosom ourselves to our friends on a subject that interests us, there is sometimes a glow of sentiment and warmth of expression that pleases, though it conveys nothing particularly ingenious or original.

The title of the book does not well express its contents. The public is too well accustomed to books that have not much correspondence with their titles, to be surprized at this. But it would have been an imposition

PREFACE.

tion of a worse kind to have changed the title in this new edition. The truth is, the subjects here treated, are so different, that it was impossible to find any title, that could fully comprehend them. Yet unconnected as they seem to be, there was a certain train of ideas that led to them, which it may not be improper to explain.

When we attend to the many advantages which Mankind possess above the inferior Animals, it is natural to enquire into the use we make of those advantages. This leads to the consideration of Man in his savage state, and through the progressive stages of human society.

Man in his savage state is, in some respects, in a worse condition than any other animal. He has indeed superior faculties, but as he does not possess, in so great a degree as other animals, the internal principle of instinct to direct these faculties to his greatest good, they are often perverted in such a manner as to render him more unhappy. He possesses bodily strength, agility, health, and

what

what are called the animal faculties, in greater perfection, than Men in the more advanced states of society; but the nobler and more distinguishing principles of human Nature lie in a great measure dormant. Like a beast of prey he passes his time generally in quest of food, or in supine sloth. He often displays the instinctive courage of a Tyger or the cunning of a Fox, though seldom tempered with that spirit of equity, generosity, and forgiveness, which alone renders courage a virtue.

There is a certain period in the progress of society, in which Mankind appear to the greatest advantage. In this period they possess the bodily powers and all the animal functions in their full vigour. They are bold, active, steady, ardent in the love of liberty and their native country. Their manners are simple, their social affections warm, and though they are much influenced by the ties of blood, yet they are generous and hospitable to strangers. Religion is universally regarded among them, though disguised by a variety of superstitions.

PREFACE. v

perftitions. This ftate * of fociety, in which Nature fhoots wild and free, encourages the high exertions of fancy and paffion, and is therefore peculiarly favourable to the arts depending on thefe; but for the fame caufe it checks the progrefs of the rational powers, which require coolnefs, accuracy, and an imagination perfectly fubdued and under the controul of reafon. The wants of Nature, likewife, being few, and eafily fupplied, require but little of the affiftance of ingenuity; though what moft effectually retards the progrefs of knowledge among fuch a people, is the difficulty of communicating and tranfmitting it from one perfon to another.

A very beautiful picture of this ftate of fociety is exhibited in the words of Offian. There we meet with Men poffeffing that high fpirit of independance, that elevation and dignity of foul, that contempt of death, that attachment to their friends and to their country, which has rendered the memory of the Greek

* Dr. Blair.

Greek and Roman Heroes immortal. But where shall we find their equals in ancient or modern story, among the most savage or the most polished nations, in those gentler virtues of the heart, that accompanied and tempered their heroism? There we see displayed the highest martial spirit, exerted only in the defence of their friends and of their country. We see there dignity without ostentation, courage without ferocity, and sensibility without weakness. Possessed of every sentiment of justice and humanity, this singular people never took those advantages, which their superior valour, or the fortune of war gave them over their enemies. Instead of massacring their prisoners in cold blood, they treated them with kindness and hospitality; they gave them the feast of shells, and, with a delicacy that would do honour to any age, endeavoured, by every art, to sooth the sense of their misfortunes, and generously restored them to their freedom. If an enemy fell in battle, his body was not insulted, nor dragged at the chariot-wheels

PREFACE. vii

wheels of the conqueror. He received the laſt honours of the warrior. The ſong of Bards aroſe. Theſe ſons of liberty were too juſt to encroach on the rights of their neighbours, and had magnanimity enough to protect the feeble and defenceleſs, inſtead of oppreſſing and enſlaving them. As they required no ſlaves to do the laborious and ſervile offices of life, they were ſtill leſs difpoſed to degrade their women to ſo mean and ſo wretched a ſituation. How humane, how noble does this conduct appear, when compared with the ungenerous treatment which women meet with among all barbarous nations, and which they ſometimes have met with among people who have been always difplayed to the world as patterns of wiſdom and virtue! There they have been condemned to the moſt miſerable ſlavery, in offices unſuitable to the delicacy of their conſtitutions, difproportionate to their ſtrength, and which muſt have totally extinguiſhed the native chearfulneſs of their ſpirits. Thus have Men inverted the order of Nature, and

taken

taken a mean and illiberal advantage of that weakneſs, of which they were the natural guardians, in order to indulge the moſt deſpicable ſloth, or to feed a ſtupid pride, which diſdained thoſe employments that Nature has made neceſſary for the ſubſiſtence and comfort of Human Life; and by this means have deſervedly cut themſelves off from the principal pleaſures of ſocial and domeſtic life. The Women deſcribed by Oſſian, have a character as ſingular as that of his Heroes. They poſſeſs the high ſpirit and dignity of Roman Matrons, united to all the ſoftneſs and delicacy ever painted in modern Romance. The hiſtory of theſe people ſeems to be juſtly referred to a period, much farther diſtant than that of chivalry; and though we make the largeſt allowance for the painting of a ſublime poetic Genius, yet we muſt ſuppoſe, that the manners and ſentiments he deſcribes had their foundation in real life, as much as thoſe deſcribed by Homer. A Poet may heighten the features and colouring of his ſubject, but if he deſerts

Nature,

PREFACE. ix

Nature, if he describes sentiments and manners unknown to his readers, and which their hearts do not recognize, it is certain he can neither be admired nor understood. The existence of such a People, in such an age and country, and of such a Poet to describe them, is one of the most extraordinary events in the history of mankind, and well deserving the attention of both philosophers and critics, especially since this is perhaps the only period where it is not only possible but easy to ascertain or disprove the reality of the fact, of which some people pretend still to doubt.—But I return to our subject.

Such a state of society as I was before describing, seldom lasts long. The power necessarily lodged in the hands of a few, for the purposes of public safety and utility, is soon abused. Ambition and all its direful consequences succeed. As the human faculties expand themselves, new inlets of gratification are discovered. The intercourse in particular with other nations brings an accession

PREFACE.

cession of new pleasures, and consequently of new wants. The advantages attending an intercourse and commerce with foreign nations are, at first view, very specious and attracting. By these means the peculiar advantages of one climate are, in some degree, communicated to another; a free and social intercourse is promoted among Mankind; knowledge is enlarged, and prejudices are removed. On the other hand, it may be said, that every country, by the help of industry, produces whatever is necessary to its own inhabitants; that the necessities of Nature are easily gratified, but the cravings of false appetite, and a deluded imagination, are endless and insatiable; that when men leave the plain road of Nature, superior knowledge and ingenuity, instead of combating a vitiated taste and inflamed passions, are employed to justify and indulge them; that the pursuits of commerce are destructive of the health and lives of the human species, and that this destruction falls principally upon

those

those who are most distinguished for their activity, spirit, and capacity.

But one of the most certain consequences of a very extended commerce, and of what is called the most advanced and polished state of society, is an universal passion for riches, which corrupts every sentiment of Taste, Nature, and Virtue. This at length reduces human Nature to the most unhappy state in which it can ever be beheld. The constitution both of body and mind becomes sickly and feeble, unable to sustain the common vicissitudes of life without sinking under them, and equally unable to enjoy its natural pleasures, because the sources of them are cut off or perverted. In this state money becomes the universal idol to which every knee bows, to which every principle of Virtue and Religion yields, and to which the health and lives of the greater part of the species are every day sacrificed. So totally does this passion pervert the human heart, that it extinguishes or conquers the natural attachment between the sexes, and in defiance of of every sentiment of Nature and sound policy, makes people look even upon their own children as an incumbrance and oppression. Neither does money, in exchange for all this, procure happiness, or even pleasure in the limited sense of the word; it yields only food for a restless, anxious, insatiable vanity, and abandons Men to dissipation, languor, disgust and misery. In this situation, patriotism is not only extinguished, but the very pretension to it is treated with ridicule: What are called public views, do not regard the encouragement of population, the promoting of virtue, or the security of liberty; they regard only the enlargement of commerce and the extension of conquest. When a nation arrives at this pitch of depravity, its duration as a free state must be very short, and can only be protracted by the accidental circumstances of the neighbouring nations being equally corrupted, or of different diseases in the state ballancing and counter-acting one another. But when once a free, an opulent and

luxurious

PREFACE. xiii

luxurious people, lose their liberty, they become of all slaves the vilest and most miserable.

We shall readily acknowledge, at the same time, that in a very advanced and polished state of society, human Nature appears in many respects to great advantage. The numerous wants which luxury creates, give exercise to the powers of invention in order to satisfy them. This encourages many of the elegant arts, and in the progress of these, some natural principles of taste, which in more simple ages lay latent in the human Mind, are awakened, and become proper and innocent sources of pleasure. The understanding likewise, when it begins to feel its own powers, expands itself, and pushes its enquiries into Nature with a success incredible to more ignorant nations. This state of society is equally favourable to the external appearance of manners, which it renders humane, gentle and polite. It is true, that these improvements are often so perverted, that they bring no accession of happiness to Mankind.

kind. In matters of taste, the great, the sublime, the pathetic, are first brought to yield to regularity and elegance; and at length are sacrificed to the most childish passion for novelty and the most extravagant caprice. The enlarged powers of understanding, instead of being applied to the useful arts of life, are dissipated upon trifles, or wasted upon impotent attempts to grasp at subjects above their reach; and politeness of manners comes to be the cloak of dissimulation. Yet still those abuses seem in some measure to be only accidental.

It was this consideration of Mankind in the progressive stages of society, that led to the idea, perhaps a very romantic one, of uniting together the peculiar advantages of these several stages, and cultivating them in such a manner as to render human life more comfortable and happy. However impossible it may be to realize this idea in large societies of Men, it is surely practicable among individuals. A person without losing any one substantial pleasure

PREFACE. xv

sure that is to be found in the moſt advanced ſtate of ſociety, but on the contrary in a greater capacity to reliſh them all, may enjoy perfect vigour of health and ſpirits; he may have the moſt enlarged underſtanding and apply it to the moſt uſeful purpoſes; he may poſſeſs all the principles of genuine Taſte, and preſerve them in their proper ſubordination; he may poſſeſs delicacy of ſentiment and ſenſibility of heart, without being a ſlave to falſe refinement or caprice. Simplicity may be united with elegance of manners; a humane and gentle temper may be found conſiſtent with the moſt ſteady and reſolute ſpirit, and religion may be revered without bigotry or enthuſiaſm.

Such was the general train of ſentiments that gave riſe to the following Treatiſe. But the reader will find it proſecuted in a very imperfect and deſultory manner. When it was firſt compoſed, the author thought himſelf at liberty to throw out his ideas without much regard to method or arrangement, and to

enlarge

enlarge more or less on particular parts of his subject, not in proportion to their importance, but as fancy at the time dictated. He would with pleasure have attempted to rectify these imperfections, which he has reason to be ashamed of in a work offered to the public; but the circumstances which he formerly mentioned put that entirely out of his power.

A

Comparative View, &c.

SECTION I.

HUMAN Nature has been confidered in very different and oppofite lights. Some have painted it in a moſt amiable form, and carefully ſhaded every weakneſs and deformity. They have reprefented vice as foreign and unnatural to the Human Mind, and have maintained that what paſſes under that name is, in general, only an exuberance of virtuous difpofitions, or good affections improperly

improperly directed, but never proceeds from any inherent malignity or depravity of the heart itself.— The Human Understanding has been thought capable of penetrating into the deepest recesses of nature, of imitating her works, and, in some cases, of acquiring a superiority over them.

Such views are generally embraced by those who have good hearts and happy tempers, who are beginning the world, and are not yet hackney'd in the ways of Men, by those who love science and have an ambition to excel in it; and they have an obvious tendency to raise the genius and mend the heart, but are the source of frequent and cruel disappointments.—

Others have represented Human Nature as a sink of depravity and wretchedness, have supposed this its natural state, and the unavoidable lot of humanity: They have represented the Human Understanding as weak and short-sighted, the Human Power as extremely feeble and limited, and have treated all attempts to

enlarge

enlarge them as vain and chimerical. —Such reprefentations are greedily adopted by Men of narrow and contracted hearts, and of very limited genius, who feel within themfelves the juftnefs of the defcription. It muft be owned however, that they are often agreeable and foothing to Men of excellent and warm affections, but of too great fenfibility of fpirit, whofe tempers have been hurt by frequent and unmerited difappointments.

A bad opinion of Human Nature readily produces a felfifh difpofition, and renders the temper cheerlefs and unfociable; a mean opinion of our intellectual faculties depreffes the genius, as it cuts off all profpect of attaining a much greater degree of knowledge than is poffeft at prefent, and of carrying into execution any grand and extenfive plans of improvement.

It is not propofed to infift further on the feveral advantages and difadvantages of thefe oppofite views of Human Nature, and on their influence in forming a character.—Perhaps

haps that View may be the safest which considers it as formed for every thing that is good and great, which sets no bounds to its capacities and powers, but looks on its present attainments as trifling and inconsiderable.

Enquiries into Human Nature, tho' of the last importance, have been prosecuted with little care and less success. This has been owing partly to the general causes which have obstructed the progress of the other branches of knowledge, and partly to the peculiar difficulties of the subject. Enquiries into the structure of the Human Body have indeed been prosecuted with great diligence and accuracy. But this was a matter of no great difficulty. It required only labour and a steady hand. The subject was permanent; the Anatomist could fix it in any position, and make what experiments on it he pleased.

The Human Mind, on the other hand, is an object extremely fleeting, not the same in any two individuals, and ever varying even in the same person.

perſon. To trace it thro' its almoſt endleſs varieties, requires the moſt profound and extenſive knowledge, and the moſt piercing and collected genius. But tho' it be a matter of great difficulty to inveſtigate and aſcertain the laws of the mental conſtitution, yet there is no reaſon to doubt, however fluctuating it may ſeem, of its being governed by laws as fixt and invariable as thoſe of the Material Syſtem.

It has been the misfortune of moſt of thoſe who have ſtudy'd the philoſophy of the Human Mind, that they have been little acquainted with the ſtructure of the Human Body, and with the laws of the Animal Oeconomy; and yet the Mind and Body are ſo intimately connected, and have ſuch a mutual influence on one another, that the conſtitution of either, examined apart, can never be thoroughly underſtood. For the ſame reaſon it has been an unſpeakable loſs to Phyſicians, that they have been ſo generally inattentive to the peculiar laws of the Mind, and to their influence on the Body.

A late celebrated profeſſor of Medicine in a neighbouring nation, who perhaps had rather a clear and methodical head, than an extenſive genius or enlarged views of Nature, wrote a Syſtem of Phyſic, wherein he ſeems to have conſidered Man entirely as a Machine, and makes a feeble and vain attempt to explain all the Phænomena of the Animal Oeconomy, by mechanical and chymical principles alone. Stahl, his cotemporary and rival, who had a more enlarged genius, and penetrated more deeply into Nature, added the conſideration of the ſentient principle, and united the philoſophy of the Human Mind with that of the Human Body: but the luxuriancy of his imagination often bewildered him, and the perplexity and obſcurity of his ſtyle occaſion his writings to be little read and leſs underſtood.

Beſides theſe, there is another cauſe which renders the knowledge of Human Nature very lame and imperfect, which we propoſe more particularly to enquire into.

Man has been ufually confidered as a Being that had no analogy to the reft of the Animal Creation. The comparative anatomy of brute Animals hath indeed been cultivated with fome attention; and hath been the fource of the moft ufeful difcoveries in the anatomy of the Human Body: But the comparative Animal Oeconomy of Mankind and other Animals, and comparative Views of their ftates and manner of life, have been little regarded. The pride of Man is alarmed, in this cafe, with too clofe a comparifon, and the dignity of philofophy will not eafily ftoop to receive a leffon from the inftinct of Brutes. But this conduct is very weak and foolifh. Nature is a whole, made up of parts, which though diftinct, are yet intimately connected with one another. This connection is fo clofe, that one fpecies often runs into another fo imperceptibly, that it is difficult to fay where the one begins and the other ends. This is particularly the cafe with the loweft of one fpecies, and the higheft of that immediately below it. On this account no one link of the great chain can be perfectly underſtood, without the knowledge, at leaſt, of the links that are neareſt to it.

In comparing the different ſpecies of Animals, we find each of them poſſeſſed of powers and faculties peculiar to themſelves, and admirably adapted to the particular ſphere of action which Providence has allotted them. But, amidſt that infinite variety which diſtinguiſhes each ſpecies, we find many qualities in which they are all ſimilar, and ſome which they have in common.

Man is evidently at the head of the Animal Creation. He ſeems not only to be poſſeſt of every ſource of pleaſure, in common with them, but of many others, to which they are altogether ſtrangers. If he is not the only Animal poſſeſt of reaſon, he has it in a degree ſo greatly ſuperior, as admits of no compariſon.

* That inſenſible gradation ſo conſpicuous in all the works of Nature, fails, in comparing Mankind with
other

* Buffon.

other Animals. There is an infinite diſtance between the faculties of a Man, and thoſe of the moſt perfect Animal; between intellectual power, and mechanic force; between order and deſign, and blind impulſe; between reflection and appetite.

One Animal governs another only by ſuperior force or cunning, nor can it by any addreſs or train of reaſoning ſecure to itſelf the protection and good offices of another. There is no ſenſe of ſuperiority or ſubordination among them*.

Their want of language ſeems owing to their having no regular train or order in their ideas, and not to any deficiency in their organs of ſpeech. Many Animals may be taught to ſpeak, but none of them can be taught to connect any ideas to the words they pronounce. The reaſon therefore, why they do not expreſs themſelves by combined and regulated ſigns, is, becauſe they have

* Inſtances from bees, birds of paſſage, and ſuch like, do not contradict this obſervation, if rightly underſtood.

have no regular combination in their ideas.

There is a remarkable uniformity in the works of Animals. Each individual of a species does the same things, and in the same manner as every other of the same species. They seem all to be actuated by one soul. On the contrary, among Mankind, every individual thinks and acts in a way almost peculiar to himself. The only exception to this uniformity of character in the different species of Animals, seems to be among those who are most connected with Mankind, particularly dogs and horses.

All Animals express pain and pleasure by cries and various motions of the body; but laughter and shedding of tears are peculiar to Mankind. They seem to be expressions of certain emotions of the soul unknown to other Animals, and are scarcely ever observed in infants till they are about six weeks old. The pleasures of the imagination, the pleasure arising from science, from the fine arts, and from the principle
of

of curiofity, are peculiar to the Human Species. But above all, they are diftinguifhed by the Moral Senfe, and the happinefs flowing from religion, and from the various intercourfes of focial life.

We propofe now to make fome obfervations on certain advantages which the lower Animals feem to poffefs above us, and afterwards to enquire how far the advantages poffeft by Mankind are cultivated by them in fuch a manner as to render them happier as well as wifer and more diftinguifhed.

There are many Animals who have fome of the external fenfes more acute than We have; fome are ftronger, fome fwifter; but thefe and fuch other qualities, however advantageous to them in their refpective fpheres of life, would be ufelefs and often very prejudicial to us. But it is a very ferious and interefting queftion, whether they poffefs not certain advantages over us, which are not the refult of their particular ftate of life, but are advantages in thofe points, where we ought

ought at least to be on a level with them.

Is it not notorious that all Animals, except ourselves, enjoy every pleasure their Natures are capable of, that they are strangers to pain and sickness, and, abstracting from external accidents, arrive at the natural period of their Being? We speak of wild Animals only. Those that are tame and under our direction partake of all our miseries.—Is it a necessary consequence of our superior faculties, that not one of ten thousand of our species dies a natural death, that we struggle through a * *frail and feverish being*, in continual danger of sickness, of pain, of dotage, and the thousand nameless ills that experience shews to be the portion of human life?—If this is found to be the designed order of Nature, it becomes us cheerfully to submit to it; but if these evils appear to be adventitious and unnatural to our constitution, it is an enquiry of the last importance, whence they arise and how they may be remedied.

There

* Milton.

There is one principle which prevails univerfally in the Brute Creation, and is the immediate fource of all their actions. This principle, which is called Inftinct, determines them by the fhorteft and moft effectual means to purfue what their feveral conftitutions render neceffary.

It feems to have been the general opinion that this principle of Inftinct was peculiar to the Brute Creation; and that Mankind were defigned by Providence, to be governed by the fuperior principle of Reafon, entirely independent of it. But a little attention will fhew, that Inftinct is a principle common to us and the whole Animal world, and that, as far as it extends, it is a fure and infallible guide; tho' the depraved and unnatural ftate, into which Mankind are plunged, often ftifles its voice, or renders it impoffible to diftinguifh it from other impulfes which are accidental and foreign to our Nature.

Reafon indeed is but a weak principle in Man, in refpect of Inftinct, and is generally a more unfafe guide.

— The

— The proper province of Reason is to investigate the causes of things, to shew us what consequences will follow from our acting in any particular way, to point out the best means of attaining an end, and, in consequence of this, to be a check upon our Instincts, our tempers, our passions, and our tastes: But these must still be the immediately impelling principles of action. In truth, life, without them, would not only be joyless and insipid, but quickly stagnate and be at an end.

Some of the advantages, which the Brute Animals have over us, are possessed in a considerable degree by those of our own species, who being but just above them, and guided in a manner entirely by Instinct, are equally strangers to the noble attainments of which their Natures are capable, and to the many miseries attendant on their more enlightened brethren of Mankind.

It is therefore of the greatest consequence, to enquire into the Instincts that are natural to Mankind, to separate them from those cravings which

which bad habits have occafioned, and, where any doubt remains on this fubject, to enquire into the analogous Inftincts of other Animals, particularly into thofe of the favage part of our own fpecies.

But a great difficulty attends this enquiry. There has never yet been found any clafs of Men who were entirely governed by Inftinct, by Nature, or by common fenfe. The moft barbarous nations differ widely in their manners from one another, and deviate as much from Nature in many particulars, as the moft polifhed and moft luxurious. They are equally guided by reafon, varioufly perverted by prejudice, cuftom, and fuperftition. Yet a difcerning eye will often be able to trace the hand of Nature where her defigns are moft oppofed, and will fometimes be furprifed with marks of fuch juft and acute reafoning among favage Nations, as might do honour to the moft enlightened. In this view the civil and natural hiftory of Mankind becomes a ftudy not merely fitted to amufe, and gratify curiofity,
but

but a study subservient to the noblest views, to the cultivation and improvement of the Human Species.

It is evident that in comparing Men with other Animals, the Analogy must fail in several respects, because they are governed solely by the unerring principle of Instinct, whereas Men are directed by other principles of action along with this, particularly by the feeble and fluctuating principle of Reason. But altho' in many particular instances it may be impossible to ascertain what is the natural and what is the artificial State of Man, to distinguish between the voice of Nature and the dictates of Caprice, and to fix the precise boundary between the provinces of Instinct and Reason; yet all Mankind agree to admit, in general, such distinctions, and to condemn certain actions as trespasses against Nature, as well as deviations from Reason. Men may dispute whether it be proper to let their beards and their nails grow, on the principle of its being natural; but every Human Creature would be shocked with the impropriety

priety of feeding an infant with Brandy inſtead of its Mother's Milk, from an inſtant feeling of its being an outrage done to Nature. In order however to avoid all altercation and ambiguity on this ſubject, we ſhall readily allow that it is our buſineſs, in the conduct of life, to follow whatever guide will lead us to the moſt perfect and laſting happineſs. We apprehend that where the voice of Nature and Inſtinct is clear and explicit, it will be found the fureſt guide, and where it is ſilent or doubtful, we imagine it would be proper to attend to the analogy of Nature among other Animals, not to be an abſolute rule for our conduct, but as a means of furniſhing light to direct it; and we admit, that, in order to determine what truly is moſt proper for us, the ultimate Appeal muſt be made to cool and impartial Experience.

We ſhould likewiſe avail ourſelves of the obſervations made on tame Animals in thoſe particulars where Art has in ſome meaſure improved upon Nature. Thus by a proper attention

attention we can preferve and improve the breed of Horfes, Dogs, Cattle, and indeed of all other Animals. Yet it is amazing that this Obfervation was never transferred to the Human Species, where it would be equally applicable. It is certain, that notwithftanding our promifcuous Marriages, many families are diftinguifhed by peculiar circumftances in their character. This Family Character, like a Family Face, will often be loft in one generation and appear again in the fucceeding. Without doubt, Education, Habit, and Emulation, may contribute greatly in many cafes to preferve it, but it will be generally found, that, independent of thefe, Nature has ftamped an original impreffion on certain Minds, which Education may greatly alter or efface, but feldom fo entirely as to prevent its traces from being feen by an accurate obferver. How a certain character or conftitution of Mind can be tranfmitted from a Parent to a Child, is a queftion of more difficulty than importance. It is indeed equally
difficult

difficult to account for the external refemblance of features, or for bodily difeafes being tranfmitted from a Parent to a Child. But we never dream of a difficulty in explaining any appearance of Nature, which is exhibited to us every day.—A proper attention to this fubject would enable us to improve not only the conftitutions, but the characters of our pofterity. Yet we every day fee very fenfible people, who are anxioufly attentive to preferve or improve the breed of their Horfes, tainting the blood of their Children, and entailing on them, not only the moft loathfome difeafes of the Body, but madnefs, folly, and the moft unworthy difpofitions, and this too when they cannot plead being ftimulated by neceffity, or impelled by paffion.

We fhall now proceed to enquire more particularly into the comparative ftate of Mankind and the inferior Animals.

By the moft accurate calculation, one half of Mankind die under eight years of age. As this mortality is
greateft

greateſt among the moſt luxurious part of Mankind, and gradually decreaſes in proportion as the diet becomes ſimpler, the exerciſe more frequent, and the general method of living more hardy, and as it doth not take place among wild Animals, the general foundations of it are ſufficiently pointed out. The extraordinary havock made by diſeaſes among Children, is owing to the unnatural treatment they meet with, which is ill ſuited to the ſingular delicacy of their tender frames. Their own Inſtincts, and the conduct of Nature in rearing other Animals, are never attended to, and they are incapable of helping themſelves. When they are farther advanced in life, the voice of Nature becomes too loud to be ſtifled, and then, in ſpite of the influence of corrupted and adventitious taſte, will be obeyed.

Though it is a maxim univerſally allowed, that a multitude of inhabitants is the firmeſt ſupport of a ſtate, yet the extraordinary mortality among Children has been little attended

tended to by Men of public spirit. It is thought a natural evil, and therefore is submitted to without examination *. But the importance of the question will justify a more particular enquiry, whether the evil be really natural and unavoidable.

It is an unpopular attempt to attack prejudices established by time and habit, and secured by the corruptions of luxurious life. It is equally unpleasant to attempt the refor-

* Thus the loss of a thousand men in an engagement arouzes the public attention, and the severest scrutiny is made into the cause of it, while the loss of thrice that number by sickness passes unregarded: yet the latter calamity is by far the most grievous, whether we regard the State, or the melancholy fate of the unhappy sufferers; and therefore calls more loudly for a Public Enquiry. Perhaps in the one case the loss was inevitable, and might lead to victory; the men faced danger with intrepidity, full of the hopes of conquest if they survived, or of dying honourably in the cause of their country. Perhaps in the other case the evil by proper management, might have been prevented: the men perished without being able to make any effort for their preservation; they saw the gradual approaches of death in all its terrors, and fell unlamented, and unsupported by that military ardor and thirst of glory which enable them to despise it in the field.

reformation of abuses, without the least prospect of success. Yet there is a secret pleasure in pleading the cause of humanity and helpless innocence.

Many reasons have been assigned, why the state of Infancy is the most sickly; and why so great a proportion of the human Species is cut off at that early period. Physicians have insisted largely on the unavoidable dangers arising from the sudden and total change of the animal Oeconomy of Infants, that commences immediately upon the Birth; and on the dangers arising from the free admission of the external air to their bodies at that time. They have expatiated on the high degree of irritability of their Nervous System, the delicacy of their whole frame, and the acescency of their food. A little reflection, however, may shew us, that this account of the matter, tho' plausible at first view, is not satisfactory. This single consideration refutes it, That all these alleged causes of the sickliness of Infants are not peculiar to the Human Species, but

but are found among many other Animals, without being attended with fuch effects; that the difeafes, moft fatal to Children, are not found among the Savage part of Mankind; and that they prevail, in exact proportion to the progrefs of Effeminacy and Luxury; and in proportion as people forfake the plain dictates, of Inftinct and Nature, to follow the Light of what they are pleafed to call Reafon.

There is, in truth, a greater luxuriancy of Life and Health in Infancy, than in any other period of Life. Infants, we acknowledge, are more delicately fenfible to Injury, than thofe advanced in Life; but, to compenfate this, their Fibres and Veffels are more capable of Diftenfion, their whole Syftem is more flexible, their Fluids are lefs acrid, and lefs difpofed to Putrefcence; they bear all Evacuations more eafily, except that of blood, and, which is an important circumftance in their favour, they never fuffer from the terrors of a diftracted Imagination. Their Spirits are lively and equal;
they

they quickly forget their paſt Sufferings, and never anticipate the future. In conſequence of theſe advantages, Children recover from diſeaſes, under ſuch unfavourable ſymptoms as are never ſurvived by Adults. If they waſte more quickly under ſickneſs, their recovery from it is quick in proportion; and generally more compleat than in older people; as diſeaſes ſeldom leave thoſe baneful effects on their Conſtitutions, ſo frequent in thoſe of Adults. In ſhort, a Phyſician ought ſcarce ever to deſpair of a Child's Life, while it continues to breathe.

Every other Animal brings forth its young without any aſſiſtance; but We judge Nature inſufficient for that work; and think a Midwife underſtands it better.—What numbers of Infants as well as of Mothers are deſtroyed by the prepoſterous management of theſe Artiſts, is well known to all who have enquired into this matter. The moſt knowing and ſucceſsful practitioners, if they are candid, will own, that in common and natural caſes, Nature is

entirely

entirely fufficient, and that their bufinefs is only to affift her efforts in cafe of weaknefs of the Mother, or an unnatural pofition of the Child.

As foon as an Infant comes into the world, our firft care is to cram it with phyfic.—There is a glareous liquor contained in the bowels of Infants and many other Animals when they are born, which it is neceffary to carry off. The medicine which Nature has prepared for this purpofe is the Mother's firft milk. This indeed anfwers the end very effectually; but we think fome drug forced down the Child's throat will do it much better. The compofition of this varies according to the fancy of the good Woman who prefides at the birth.—It deferves to be remarked, when we are on this fubject, that calves, which are the only Animals generally taken under our peculiar care in thefe circumftances, are treated in the fame manner. They have the fame fort of phyfic adminiftered to them, and often with the fame fuccefs; many of them dying under the operation, or of its confequences:

and we have the greatest reason to think that more of this species of Animals die at this period, than of all the other species of Animals we see in these circumstances, put together, our own only excepted.

Notwithstanding the many moving calls of natural Instinct in the Child to suck the Mother's breast, yet the usual practice has been, obstinately to deny that indulgence till the third day after the birth. By this time the suppression of the natural evacuation of the milk, usually bringing on a fever, the consequence proves often fatal to the Mother, or puts it out of her power to suckle her Child at that time. The sudden swelling of the breasts, which commonly happens about the third day, is another bad consequence of this delay. When the breasts become thus suddenly and greatly distended, a child is not only utterly unable to suck, but, by its cries and struggling, fatigues and heats, both itself and the Mother. This is another frequent cause, which prevents nursing. — We must observe here, to the ho-

nor

nor of the gentlemen who had the care of the lying-in hofpital in London, that they were the firſt who, in this inſtance, brought us back to Nature and common fenfe; and by this means have preferved the lives of thoufands of their fellow-creatures. They ordered the Children to be put to the Mother's breaſt as foon as they fhewed a defire for it, which was generally within ten or twelve hours after the birth. This rendered the ufual dofe of phyfic unneceffary, the milk-fever was prevented, the milk flowed gradually and eafily into the breaſts, which before were apparently empty, and things went fmoothly on in the natural way. We are forry however to obferve, that this practice is not likely to become foon general. Phyficians do not concern themfelves with fubjects of this kind, nor with the regimen of Mankind, unlefs their advice is particularly afked. Thefe matters are founded on eſtabliſhed cuſtoms and prejudices, which it is difficult to conquer, and dangerous to attack; nor will it ever be attempted

tempted by Men who depend on the favor and caprice of the world for their fubfiftence, and who find it their intereft rather to footh prejudices than to oppofe them. If a Mother therefore is determined not to nurfe her own Infant, fhe fhould, for her own fake, fuckle it at leaft three or four weeks, and then wean it by degrees from her own breaft. In this way the more immediate danger arifing from repelling the milk, is prevented.

When a Mother does not nurfe her own Infant, fhe does open violence to Nature; a violence unknown among all the inferior Animals, whom Nature intended to fuckle their young: unknown among the moft barbarous nations; and equally unknown among the moft polifhed, in the pureft ages of Greece and Rome. The fudden check given to the great natural evacuation of Milk, at a time when her weakly ftate renders her unable to fuftain fo violent a fhock, is often of the worft confequence to herfelf; and the lofs to the Child is much greater than is commonly apprehended.

prehended. A Woman in this cafe runs an immediate rifk of her life by a milk-fever, befides the danger of fwelling and impofthumes of the breaft, and fuch obftructions in them as often lay the foundation of a future cancer.—Of 4,400 Women in the lying-in hofpital, only four had milk fores, and thefe had either no nipples, or former fore breafts*.

Some Women indeed have it not in their power to nurfe their Children, for want of milk; and fometimes it is equally improper for the Mother and the Child, on account of fome particular diforder which the Mother labours under. But this is very feldom the cafe. On the contrary, there are many diforders incident to Women, of which nurfing is the moft effectual cure; and delicate conftitutions are generally ftrengthened by it. In proof of this we may obferve, that while a Mother nurfes her Child, her complexion becomes clearer and more blooming, her fpirits are more uniformly chearful, her appetite is better, and her

* Nelfon.

her general habit of body fuller and ſtronger. And it is particularly worthy of obſervation, that fewer Women die while they are nurſing than at any equal period of their lives, if we except the time of pregnancy, during which it is unuſual for a Woman to die of any diſeaſe, unleſs occaſioned by ſome violent external injury.

Another great inconveniency attending the neglect of nurſing, is the depriving Women of that interval of reſpite and eaſe which Nature intended for them between Child-bearings. A woman who does not nurſe, has naturally a Child every year; this quickly exhauſts the conſtitution, and brings on the infirmities of old-age before their time; and as this neglect is moſt frequent among Women of faſhion, the delicacy of their conſtitutions is particularly unable to ſuſtain ſuch a violence to Nature. A Woman who nurſes her Child, has an interval of a year and a half or two years betwixt her Children, in which the conſtitution has time to recover its vigor*.

We may reckon, among the diſadvantages conſequent on the neglect of nurſing, the Mother's being deprived of a very high pleaſure, of the moſt tender and endearing kind, which remarkably ſtrengthens her attachment to the Infant. It is not neceſſary here to enquire into the cauſe of this particular affection which a Mother feels for the Child ſhe has ſuckled, ſuperior to that which ſhe feels for a Child ſuckled by a ſtranger; but the fact itſelf is indiſputable.

It is not eaſy to eſtimate the injury Children ſuſtain by being deprived of their natural nouriſhment, and, inſtead of it, being ſuckled by the milk of Women of different ages and conſtitutions from their Mothers. Thus far is certain, that a greater number of thoſe Children die who are nurſed by ſtrangers, than

of

* When the natural evacuation of milk from the breaſts is ſuppreſſed, it renders the diſcharge of the Lochia more copious, and of longer duration than Nature intended, which is a frequent ſource of the Fluor albus.

of thofe who are fuckled by their own Mothers. This is partly owing however to the want of that care and attention which the helplefs ftate of Infancy fo much requires, and which the anxious affection of a Mother can alone fupply. Indeed if it was not that Nurfes naturally contract a large portion of the inftinctive fondnefs of a Mother, for the Infants they fuckle, many more of them would perifh by want of care. But it fhould be obferved, that this acquired attachment cannot reafonably be expected among Nurfes, in large cities. The fame perverfion of nature and manners which prevails there among Women of fafhion, and makes them decline this duty, extends equally to thofe of lower rank: and it cannot be fuppofed that what the call of Nature, not to fpeak of love for the hufband, is unable to effectuate in the Mother, will be found in a hireling, who for a little money turns her own Infant out of doors. But tho' it is true that a Nurfe may acquire by degrees the folicitude and tendernefs of a

Mother,

Mother, yet as this takes place flowly, and only in proportion as habit takes the place of Nature, the neglected Child may perish in the mean time. There results even from this poffible advantage, an inconvenience which is itfelf fufficient to deter a Woman of any fenfibility from permitting her Infant to be fuckled by another: and this is, to have a ftranger partaking with, or rather alienating from her the rights of a Mother; to fee her Child love another Woman as well, or better than herfelf; to perceive the affection it retains for its natural parent a matter of favour, and that of its adopted one a duty: for is not the attachment of the Child the reward due to the tender cares of a Mother*? The many loathfome difeafes to which the lower clafs of Women in large cities are fubjected, is another reafon againft their being intrufted with fuch an office; difeafes which are often fatal to their little charges, or which taint their blood in a manner that

* Rouffeau.

that they and their succeeding families may feel very severely.

Children should be suckled from nine to twelve months. There are several circumstances that may point out the propriety of weaning them about that time: in many parts of Europe, and in all the Levant, Children taste nothing but their Mother's milk till they are a year old, which in general is a good rule. The call of Nature should be waited for to feed them with any thing more substantial. Many diforders are incident to Infants, by forcing other food upon them besides their Nurses milk. When we neglect the plain dictates of Instinct in this case, we cannot move a step without danger of erring, in regard to the quantity or quality of their food, or the proper times of giving it. New-born Infants are particularly apt to suffer from being stuffed with water-gruel, milk and water, weak wine whey, and other things of the like kind, which are thought perfectly mild and innocent. But the case is, Nature at this time requires very little food,
but

but a great deal of rest, as Infants sleep almost their whole time, for several weeks after they are born. When therefore something or other is continually pouring down their throats, their natural repose is interrupted, and the effects are flatulency, gripes, and all the other consequences of indigestion. It is proper to wean Children by degrees, and to make this and every subsequent alteration in their diet as gradual as possible, because too sudden transitions in this respect are often attended with the worst consequences.

While an Infant is fed by the Mother's milk alone, it may be allowed to suck as often as it pleases. It is then under the peculiar protection of Nature, who will not neglect her charge; and in this case has wisely provided against any inconvenience that may arise from the stomach being overcharged with too much milk, by making the Child throw up the superfluous quantity; which it does without sickness or straining.

If a Mother cannot or will not suckle her own Child, it should be
given

given to a Nurſe newly delivered, whoſe conſtitution both of body and of mind reſembles the Mother's as nearly as poſſible, provided that conſtitution be a good one. The Nurſe ſhould continue to live in every reſpect as ſhe has been accuſtomed to do. A tranſition from a plain diet conſiſting moſtly of vegetables, from a pure air and daily exerciſe, if not hard labour, to a full diet of animal food, fermented liquor, the cloſe air of a town, and a total want of exerciſe, cannot fail to affect the health both of the Nurſe and the Child.

The attempt to bring up an Infant entirely by the ſpoon is offering ſuch a violence to Nature, as nothing but the moſt extreme neceſſity can juſtify. If a Child was to be nouriſhed in this way, even by its Mother's milk alone, it would not anſwer. The action of ſucking, like that of chewing, occaſions the ſecretion of a liquor in the Child's mouth, which being intimately mixed with the milk, makes it ſit eaſy upon, and properly digeſt in the ſtomach.

Beſides

Besides these, there are other circumstances in the rearing of Children, in which, we apprehend, neither Instinct, nor the Analogy of Nature is properly regarded.

All young Animals naturally delight in the open air, and in perpetual motion: But we signify our disapprobation of this intention of Nature, by confining our Infants mostly within doors, and swathing them from the time they are born as tightly as possible.—This natural Instinct appears very strong when we see a Child released from its confinement, in the short interval between pulling off its day cloaths, and swathing it again before it is put to sleep. The evident tokens of delight which the little creature shews in recovering the free use of its limbs, and the strong reluctance it discovers to be again remitted to its bondage, one should imagine would strike a conviction of the cruelty and absurdity of this practice, into the most stupid of Mankind. This confinement, Boys, in general, are sooner released from; but the fairer part
of

of the Species fuffer it, in fome degree, during life.

Some nations have fancied that Nature did not give a good fhape to the head, and thought it would be better to mould it into the form of a fugar-loaf. The Chinefe think a Woman's foot much handfomer if fqueezed into a third part of its natural fize. Some African nations have a like quarrel with the fhape of the nofe, which they think ought to be laid as flat as poffible with the face. We laugh at the folly and are fhocked with the cruelty of thefe barbarians; but think, with equal abfurdity, that the natural fhape of a Woman's cheft is not fo elegant, as we can make it by the confinement of Stays.— The common effects of this practice are diforders in the ftomach and obftructions in the lungs, from their not having fufficient room to play, which, befides tainting the breath, cuts off numbers of young Women by confumptions in the very bloom of life.— But Nature has fhewn her refentment of this practice in the moft ftriking manner, by

rendering

rendering above half the Women of fashion deformed in some degree or other. Deformity is peculiar to the civilized part of Mankind, and is almost always the work of our own hands. The Turkish and Asiatic Women, who are distinguished for the elegance of their form, and the gracefulness of their carriage, are accustomed from their Infancy to wear no dress but what is perfectly loose.—The superior strength, just proportions, and agility of Savages are entirely the effects of their hardy education, of their living mostly abroad in the open air, and of their limbs never having suffered any confinement.—The Siamese, Japonese, Indians, Negroes, Savages of Canada, Virginia, Brazil, and most of the inhabitants of South America, do not swathe their Children, but lay them in a kind of large cradle lined and covered with skins or furs. Here they have the free use of their limbs; which they improve so well, that in two or three months they crawl about on their hands and knees, and in less than a year walk without any assistance.

ance. Where Children are swathed, or so closely pinioned down in their cradles, that they cannot move, the *impulsive force of the internal parts of the body disposed to increase, finds an insurmountable obstacle to the movements required to accelerate their growth The Infant is continually making fruitless efforts, which waste its powers or retard their progress. It is scarcely possible to swathe Children in such a manner as not to give them some pain; and the constant endeavour to relieve themselves from an uneasy posture, is a frequent cause of deformity. When the swathing is tight, it impedes the breathing, and the free circulation of the blood, disturbs the natural secretions, and disorders the constitution in a variety of ways. If an Infant is pinioned down in its cradle in such a manner as to prevent the superfluous humour secreted in the mouth from being freely discharged, it must fall down into the stomach; where it occasions various disorders, especially in time of teething,

* Rousseau.

teething, when there is always a very great fecretion of this fluid. Another inconvenience which attends this unnatural confinement of Children, is the keeping them from their natural action and exercife, which both retards their growth, and diminifhes the ftrength of their bodies. It is pretended that Children left thus at liberty, would often throw themfelves into poftures deftructive of the perfect conformation of their body. But if a Child ever gets into a wrong fituation, the uneafinefs it feels foon induces it to change its pofture. Befides, in thofe countries where no fuch precautions are taken, the Children are all robuft and well proportioned. It is likewife faid, that if Children were left to the free ufe of their limbs, their reftleffnefs would fubject them to many external injuries; but tho' they are * heavy, they are proportionably feeble, and cannot move with fufficient force to hurt themfelves. The true fource, however, of that wretched flavery to which they are condemned is

* Rouffeau.

is this; an Infant whose limbs are at liberty must be constantly watched, but when it is fast bound, it requires little attendance from its Nurse, and may be thrown into any corner.

It is of the utmost consequence to the health of Infants, to keep them perfectly clean and sweet. The inhabitants of the * Eastern countries, particularly Turkey, and the natives of America, are extremely attentive to this article. The confined dress of our Infants renders a great degree of attention to cleanliness peculiarly necessary. The close application of any thing acrid to the delicate and sensible skin of an Infant, gives a very speedy irritation, and is one of the most frequent causes of Children's crying.

Children when very young never cry but from pain or sickness, and therefore the cause of their distress should be accurately enquired into. If it is allowed to continue, it disturbs all the animal functions, especially the digestive powers; and from the disorders of these most of the diseases

* Buffon.

diseases incident to Children proceed. The cries of an Infant are the voice of Nature supplicating relief. It can express its wants by no other language. Instead of hearkening to this voice, we often stifle it, by putting the little wretch into a cradle, where the noise and violent motion confound all its senses, and extinguish all feelings of pain in a forced and unnatural sleep. Sometimes they are allowed to cry till their strength is exhausted. But their violent struggles to get relief, and the agitations of their passions, equally disorder their constitutions; and when a Child's first sensations partake so much of pain and distress, and when the turbulent passions are so early awaked and exercised, there is some reason to suspect they may have an influence on the future temper.

Children require a great deal of sleep, particularly in early infancy, nor should it ever be denied them. If they are allowed to be in constant motion when they are awake, which they always choose to be, there will be be no occafion for rocking them in a cradle: but the fleep which is forced, by exhaufted Nature finking to reft after fevere fits of crying, is often too long and too profound. Rocking in cradles is improper in every refpect, from the confinement they occafion, from their overheating Infants, from their difordering the digeftion of their food, and from their procuring an unnatural and forced fleep.

As Children naturally turn their eyes to the light, their beds or cradles fhould be lighted from the feet, in fuch a way as that both eyes may be equally expofed to it. If the light is on one fide, the eye that is moft frequently directed to it will become ftrongeft. This is likewife a frequent caufe of fquinting[*].

The mifmanagement of Children is principally owing to over-feeding, over-clothing, want of exercife, and of frefh air[†]. Though, as was before obferved, a young Child never cries but

---

[*] Buffon.
[†] See a very fpirited and judicious effay on Nurfing, by Dr. Cadogan.

but from pain or ſickneſs, yet the univerſal remedy abſurdly applied for all its diſtreſſes, is giving it ſomething to eat or to drink, or rocking it in a cradle. If the wants and motions of a child are attended to, it will be found to ſhew ſeveral ſigns of deſiring food before it cries for it, the firſt ſenſations of hunger never being attended with pain. Indeed theſe ſigns are ſeldom obſerved, becauſe Children are ſeldom ſuffered to be hungry. If they were regularly fed only thrice a day, at ſtated intervals, after they are weaned, the ſignals of returning hunger would be as intelligible as if they ſpoke; but while they are crammed with ſome traſh every hour, the calls of natural appetite can never be heard. Their food ſhould be ſimple, and of eaſy digeſtion, and ſhould never be taken hot: after they are weaned, till they are three years old, it ſhould conſiſt of plain milk, panada, well-fermented bread, barley-meal porridge; and at dinner plain light broth with barley or rice. All kinds of paſtry, puddings, cuſtards, &c. where
the

the chief ingredients are unfermented flour, eggs, and butter, tho' generally thought to be light, lie much heavier on the ſtomach than many kinds of animal food. Fermented liquors of every kind, and all ſorts of ſpiceries, are improper. They give a ſtimulus to the digeſtive powers, which they do not require, and, by exciting a falſe appetite, are often the cauſe of their being overcharged. Their drink ſhould be pure water. The quantity of Children's food ſhould be regulated by their appetite; and as they always eat with ſome eagerneſs full as much as they ought, whenever that eagerneſs ceaſes, their food ſhould be immediately withdrawn.

The practice of putting many clothes on Children, indulging them in ſitting over the fire, ſleeping in ſmall and warm rooms, and preſerving them from being expoſed to the various inclemencies of the weather, relaxes their bodies, and enervates their minds. If Children, together with ſuch an effeminate education, are pampered with animal food,

food, rich fauces, and such other diet as overcharges their digestive powers, they become sickly as well as weak.

It is a general error, that a new-born Infant cannot be kept too warm. From this unfortunate prejudice, a healthy Child is soon made so tender, that it cannot bear the fresh air without catching cold. A Child can neither be kept too cool, nor too loose in its dress. It wants less clothing, in proportion, than a grown person, because it is naturally warmer; at least more uniformly and equally warm. This is universal among all animals. There are numberless instances of Infants, exposed and deserted, that have lived several days, in such severe weather as would have killed most adults. Many of the diseases incident to new-born Infants, and to Lying-in Women, arise from the hot regimen to which they are subjected. It is generally thought necessary to keep Lying-in Women in a constant, extorted Sweat, by confining them for several days closely to bed, in warm rooms,

rooms, where great care is taken to exclude the freſh air; by giving them all their drink warm, and obliging them to take down a larger quantity of it than their thirſt demands. If all theſe methods prove inſufficient to force out the deſired Sweat, the aſſiſtance of ſudorific medicines, ſometimes of the heating kind, is called in. There is the greateſt reaſon to believe, that the whole of this artificial Syſtem of management is highly pernicious. It is contrary to the Analogy of Nature among all other Animals, and among the uncultivated part of the human ſpecies, who, unleſs in ſome very extraordinary caſes, recover eaſily and ſpeedily, after bringing forth their young, without requiring to be kept warmer than uſual. The frequent deaths, and the ſlow and difficult recoveries of Women after Child-birth, ſhew plainly that there is an error ſomewhere. It is the refuge of ignorance, or the blindneſs of prejudice, to ſay, that theſe evils are natural and unavoidable. The Conſtitution of a Lying-in Woman is indeed naturally

more

more irritable than usual, but this irritability is much increased by a hot regimen, and by keeping her constantly dissolved in Sweats: the effect of which is, to weaken her so much, that the least application of external cold often produces the most dangerous consequences. This is considered as an additional reason for keeping the unhappy Woman still warmer. It generally happens, that a woman, for some days after her delivery, has a constant Moisture on her Skin; this natural Moisture is most effectually promoted by keeping her as cool as in her usual health. If the heat is increased, instead of this salutary Perspiration, a Fever is probably produced, which either suppresses it entirely, or is attended with a profuse colliquative Sweat; and often, in consequence of such Sweat, with a Miliary Eruption. By another fatal error, in mistaking an Effect for a Cause, this Miliary Eruption is considered as a critical and highly salutary translation of some imaginary morbid matter to the Skin; which ought to be promoted,

moted, by a warm regimen and sudorific medicines. Thus, by leaving the plain road of Nature and common Sense, people involve themselves in a labyrinth of errors, and fancy they are curing Diseases, when, in truth, they are creating them. It is a certain fact, however strange it may appear, that in a well-regulated Lying-in Hospital, Women recover sooner, and are subjected to fewer accidents after Child-birth, notwithstanding the unavoidable exposure to more light and noise, than Ladies of Fashion, who are thought to possess every possible conveniency, in their own houses. The reason is obvious: In such an Hospital, the Women lie in a large ward, kept cool and well ventilated, and under the direction and absolute government of a Physician, who is not fettered by other people's prejudices, but feels himself at full liberty to act according to the dictates of his own Understanding and Experience.

But we return to our Subject.—Children should have no shoes or stockings, at least till they are able to to run abroad. They would ſtand firmer, learn to walk ſooner, and have their limbs better proportioned, if they were never cramped with ligatures of any kind. Beſides, ſtockings are a very uncleanly piece of dreſs, and always keep an Infant's legs cold and wet, if they are not ſhifted almoſt every hour.

The active principle is ſo vigorous and overflowing in a Child, that it loves to be in perpetual motion itſelf, and to have every object around it in motion. This exuberant activity is given it for the wiſeſt purpoſes; as it has more to do and more to learn in the firſt three years of its life, than it has in thirty years of any future period of it. But that lively and reſtleſs ſpirit, which in infancy ſeemed to animate every thing around it, gradually contracts itſelf, as the Child advances in Life, nature requiring no more motion than is neceſſary for its preſervation, and ſinks at laſt into that calm and ſtillneſs which cloſe the latter days of human life.

We should freely indulge this active spirit and the restless curiosity of Children, by allowing them to move about at their pleasure. This exercise gives strength and agility to their limbs and vigour to their constitutions. They should be allowed and even encouraged to handle objects from their earliest infancy, and be suffered to approach them as soon as they are able to move on their hands and knees. It is only by touch that we acquire just ideas of the figure and situation of bodies, and therefore we cannot be too early accustomed to examine by this sense every visible body within our reach. All these purposes, however, are frustrated by Infants being confined in their Nurses arms till they are able to walk alone. This confinement is likewise very apt to give a twist to their shape, if the Nurse is not particularly careful to carry them alternately in both arms, tho' this twist may not appear for many years after. But a still more important injury may be done to them by this practice, so universal among

those

thofe of better rank; the injury arifing from their having too much or too little exercife, or from its being given them at an improper time. If a Child is fuffered to move about at its pleafure, like any other young animal, from the time it is two or three months old, unerring Inftinct will direct it to take precifely the Quantity of Exercife, and to take it at the precife times which are moft proper. But if it is carried always in a Nurfe's arms, thefe important circumftances muft be regulated by her peculiar temper or caprice. It is eafy to forefee fome of the numerous inconveniencies that muft arife from this.

Neither ought Children to be affifted, in their learning to walk, by leading-ftrings. The only ufe of thefe is to fave trouble to Nurfes, who, by allowing the Children to fwing in them, often hurt their fhape, and retard their progrefs in walking. They are lefs fubject to fall when they have no fuch artificial affiftance to depend on; and they cannot too early be made fenfible that

that they are never to expect a support or assistance in doing any thing which they are able to do for themselves. When Infants have escaped from the hands of their Nurses and are able to run about and shift for themselves, they generally do well. It is commonly thought that weakly Children should not be put on their legs, especially if they are the least bent or crooked: but experience shews that crooked legs will grow in time strong and strait by frequent walking, while disuse makes them worse and worse every day*.

Cities are the graves of the human species †. They would perish in a few generations, if they were not constantly recruited from the country. The confined, putrid air which most of their inhabitants breathe, their foul feeding, their want of natural exercise, but, above all, their debauchery, shorten their lives, ruin their constitutions, and produce a puny and diseased race of Children.

Every circumstance points out the country as the proper place for the education

* Cadogan.   † Rousseau.

education of Children; the purity of the air, the variety of ruſtic ſports, the plainneſs of diet, the ſimplicity and innocence of manners, all concur to recommend it. Crowding Children together in hoſpitals is extremely pernicious to their health, both from the confinement they are ſubjected to, and from the unwholeſome air occaſioned by a number of people living in the ſame houſe. But it is ſtill more pernicious to confine them, before they have attained their full growth and ſtrength, to ſedentary employments, where they breathe a putrid air, and are reſtrained from the free uſe of their limbs. The uſual effect of this confinement is, either to cut them off early in life, or to render their conſtitutions weak and ſickly. The inſatiable thirſt for money, not only hardens the heart againſt every ſentiment of humanity, but makes men blind to that very intereſt which they ſo anxiouſly purſue. The ſame principles of ſound policy, which induces them to ſpare their horſes and cattle, till they arrive at their full ſize and vigour,

gour, fhould naturally lead them to grant a like refpite to their Children.

Tho' diet demands the greateft attention, in puny conftitutions, yet it admits of a very great latitude in Children hardened by exercife and daily expofed to the viciffitudes of the weather. It is impoffible to afcertain what the human body may be brought to bear, if it is gradually inured to the intemperance of feafons and elements, to hunger, thirft, and fatigue. Before it hath acquired fettled habits, we may induce almoft any we pleafe, without danger; when it is once arrived at its full growth and confiftence, every material alteration is dangerous. But the delicacy and luxury of modern education deftroy the foundation of this native vigor and flexibility. Notwithftanding the variety of abfurd and unnatural cuftoms that prevail among barbarous nations, they are not fickly as we are, becaufe the hardinefs of their conftitutions enables them to bear all exceffes. The women who inhabit the ifthmus of America are plunged in cold water,
along

along with their Infants, immediately after their delivery, without any bad confequence. All thofe difeafes which arife from catching of cold, or a fudden check given to the perfpiration, are found only among the civilized part of Mankind. An old Roman or an Indian, in the purfuits of war or hunting, would plunge into a river whilft in a profufe fweat, without fear and without danger. A fimilar hardy education would make us all equally proof againft the bad effects of fuch accidents.— The greater care we take to prevent catching cold, by the various contrivances of modern luxury, the more we become fubjected to it. — We can guard againft cold only by rendering ourfelves fuperior to its influence.— There is a ftriking proof of this in the vigorous conftitutions of Children braced by the daily ufe of the cold bath; and ftill a ftronger proof, in thofe Children who are thinly clad, and fuffered to be without ftockings or fhoes in all feafons and weathers.

Nature never made any country too cold for its own inhabitants.— In cold climates she has made exercise and even fatigue habitual to them, not only from the neceffity of their fituation, but from choice, their natural diverfions being all of the athletic and violent kind. But the foftnefs and effeminacy of modern manners has both deprived us of our natural defence againft the difeafes moft incident to our own climate, and fubjected us to all the inconveniencies of a warm one, particularly to that debility and morbid fenfibility of the nervous fyftem, which lays the foundation of moft of our difeafes, and deprives us at the fame time of the fpirit and refolution to fupport them.

Moft of thofe Children who die under two years of age, are cut off by the confequences of teething. This is reckoned a natural and inevitable evil; but as all other animals, and the uncultivated part of Mankind, get their teeth without danger, there is reafon to fufpect this is not a natural evil. The procefs of Nature

ture in breeding teeth is different from her usual method of operating in the human body, which is without pain, and commonly without exciting any particular sensation. But though cutting of the teeth may be naturally attended with some pain, and even a small degree of fever, yet if a Child's constitution be perfectly sound and vigorous, probably neither of these would be followed by any bad consequence. The irritability of the nervous System, and the inflammatory disposition of the habit at this period, are probably owing in a great measure to too full living, to the constitution being debilitated by the want of proper Exercise, by the want of free Exposure to the open Air, and the numberless other Effeminacies of modern Education. Other animals facilitate the cutting of their teeth by gnawing such bodies as their gums can make some impression on. An Infant, by the same mechanical Instinct, begins very early to carry every thing to its mouth. As soon as this indication of Nature is observed, it should

be

be diligently followed, by giving the Child something to gnaw, which is inoffensive, which is cooling, and which yields a little to the pressure of its gums, as liquorice-root, hard biscuit, wax candle, and such like. A perfectly hard body, such as coral, does not answer the purpose, nor will a Child use it, when its gums are in the least pained.

We cannot help observing here, the very great prejudice which Children of better rank often sustain, by a too early application to different branches of education. The most important possession that can be secured to a Child, is a healthy and vigorous constitution, a chearful temper, and a good heart. Most sickly Children either die very soon, or drag out an unhappy life, burdensome to themselves, and useless to the public. There is nothing indeed to hinder a Child from acquiring every useful branch of knowledge, and every elegant accomplishment suited to his age, without impairing his constitution; but then the greatest attention must be had to the

powers

powers of his body and mind, that they neither be allowed to languish for want of exercife, nor be exerted beyond what they can bear. Nature brings all her works to perfection by a gradual procefs. Man, the laſt and moſt perfect of her works below, arrives at his by a very flow procefs. In the early period of life, Nature feems particularly folicitous to increafe and invigorate the bodily powers. One of the principal inſtruments ſhe ufes for this purpofe is, that reſtlefs activity which makes a Child delight to be in perpetual motion. The faculties of the mind difclofe themfelves in a certain regular fucceffion. The powers of imagination firſt begin to appear by an unbounded curiofity, a love of what is great, furpriſing, and marvellous, and, in many cafes, of what is ridiculous. The perception of what is beautiful in Nature does not come fo early. The progrefs of the affections is flower: at firſt they are moſtly of the felfiſh kind, but, by degrees, the heart dilates, and the focial and public affections make their appearance.

The

The progress of reason is extremely slow. In childhood the mind can attend to nothing but what keeps its active powers in constant agitation, nor can it take in all the little discriminating circumstances which are necessary to the forming a true judgment either of persons or things. For this cause it is very little capable of entering into abstract reasoning of any kind, till towards the age of manhood. It is even long after this period before any justness of taste can be acquired, because that requires the most improved use of the affections, of the reasoning faculty, and of the powers of imagination. If this is the order and plan of Nature in bringing Man to the perfection of his kind, it should be the business of education religiously to follow it, to assist the successive openings of the human powers, to give them their proper exercise, but to take care that they never be overcharged. If no regard is had to this rule, we may indeed accelerate the seeming maturity of our faculties, as we can rear a plant in a hot-bed, but
we

we shall never be able to bring them to that full maturity, which a more strict attention to Nature would have brought them to. This is, however, so little observed in the education of Children of better fashion, that Nature is, almost from the beginning, thwarted in all her motions. Many hours are spent every day in studies painfully disagreeable, that give exercise to no faculty but the memory, and only load it with what will probably never turn to either future pleasure or utility. Some of the faculties are over-strained, by putting them upon exertions disproportioned to their strength; others languish for want of being exercised at all. No knowledge or improvement is here acquired by the free and spontaneous exertion of the natural powers: it is all artificial and forced. Thus health is often sacrificed, by the body being deprived of its requisite exercise, the temper hurt by frequent contradiction, and the vigour of the mind impaired by unnatural and overstrained exertions. The happiest period of Human Life, the days of

health,

health, chearfulnefs and innocence, on which we always reflect with pleafure, not without fome mixture of regret, are fpent in the midft of tears, punifhments, and flavery; and this is to anfwer no other end but to make a Child a Man fome years before Nature intended he fhould be one. It is not meant here to infinuate, that Children fhould be left to form themfelves without any direction or affiftance. On the contrary, they need the moft watchful attention from their earlieft infancy, and often contract fuch bad health, fuch bad tempers, and fuch bad habits, before they are thought proper fubjects of education, as will remain with them, in fpite of all future care, as long as they live. We only intended to point out the impropriety of precipitating education, by forfaking the order in which Nature unfolds the human powers, and by facrificing prefent happinefs to uncertain futurity. There is a kind of culture that will produce a Man at fifteen, with his character and manners perfectly formed: but then

he

he is a little Man; his faculties are cramped, and he is incapable of further improvement. By a different culture he might not perhaps arrive at full maturity till five-and-twenty; but then he would be by far the superior man, bold, active, and vigorous, with all his powers capable of still further enlargement. The bufinefs of education is indeed, in every view, a very difficult tafk. It requires an intimate knowledge of Nature, as well as great addrefs, to direct a Child, before he is able to direct himfelf, to lead him without his being confcious of it, and to fecure the moft implicit obedience, without his feeling himfelf to be a flave. It requires befides fuch a conftant watchfulnefs, fuch inflexible fteadinefs, and, at the fame time, fo much patience, tendernefs, and affection, as can fcarcely be expected but from the heart of a parent.

Thefe few obfervations are felected from a great number that might be mentioned, to prove that many of the calamities complained of as peculiarly affecting the Human Species,

are

are not neceſſary conſequences of our conſtitution, but are entirely the reſult of our own caprice and folly, in paying greater regard to vague and ſhallow reaſonings, than to the plain dictates of Nature, and the analogous conſtitutions of other Animals.—They are taken from that period of life, where Inſtinct is the only active principle of our Nature, and conſequently where the analogy between us and other Animals will be found moſt compleat.—When our ſuperior and more diſtinguiſhing faculties begin to expand themſelves, the analogy becomes indeed leſs perfect. But, if we would enquire into the cauſe of our weak and ſickly habits, we muſt go back to the ſtate of Infancy. The foundation of the evil is laid there. Habit ſoon ſucceeds in the place of Nature, and, however unworthy a ſucceſſor, requires almoſt equal attention. As years advance, additional cauſes of theſe evils are continually taking place, and diſorders of the body and mind mutually inflame each other.—But this opens a field

too

too extenfive for this place. We fhall only obferve, that the decline of Human Life exhibits generally a fcene quite fingular in Nature.— The gradual decay of the more humane and generous feelings of the heart, as well as of all our boafted fuperior powers of imagination and underftanding, till at laft they are utterly obliterated, and leave us in a more helplefs and wretched fituation than that of any animal whatever, is furely of all others the moft humbling confideration to the pride of man.— Yet there is great reafon to believe that this melancholy Exit is not our natural one, but that it is owing to caufes foreign and adventitious to our Nature.— There is the higheft probability, at leaft, that if we led natural lives, we fhould retain to the laft the full exercife of all our fenfes, and the full poffeffion of thofe fuperior faculties, which we hope we fhall retain in a future and more perfect ftate of exiftence.— There is no reafon to doubt but it is in the power of art to protract life

even

even beyond the period which Nature has affigned to it. But this enquiry, however important, is trifling, when compared to that which leads us to the means of enjoying it, whilft we do live.

SECTION II.

THE advantages, which Mankind poffefs above the reft of the Animal Creation, are principally derived from Reafon, from the Social Principle, from Tafte, and from Religion. We fhall proceed to enquire how much each of thefe contribute to make life more happy and comfortable.

Reafon, of itfelf, cannot, any more than riches, be reckoned an immediate bleffing to Mankind. It is only the proper application of it, to render them more happy, that can entitle it to that name. Nature has furnifhed us with a variety of internal Senfes and Taftes, unknown to other Animals. All thefe, if properly cultivated, are fources of pleafure, but without culture, moft of them are fo faint and languid, that they

they convey no gratification to the Mind. This culture is the peculiar province of Reafon. It belongs to reafon to analyze our Taftes and Pleafures, and, after a proper arrangement of them according to their different degrees of excellence, to affign to each that degree of cultivation and indulgence which its rank deferves, and no more. But if Reafon, inftead of thus doing juftice to the various gifts of Providence, be unattentive to her charge, or beftow her whole attention on One, neglecting the reft, and if, in confequence of this, little happinefs be enjoyed in life, in fuch a cafe Reafon can with no great propriety be called a bleffing. Let us then examine its effects among thofe who poffefs it in the moft eminent degree.

The natural advantages of Genius, and a fuperior Underftanding, are extremely obvious. One unacquainted with the real ftate of human affairs, would never doubt of their fecuring to their poffeffors the moft honourable and important ftations among mankind, nor fufpect that they could ever

ever fail to place them at the head of all the useful arts and profeffions. If he were told this was not the cafe, he would conclude it muft be owing to the folly or wickednefs of Mankind, or to fome unhappy concurrence of accidents, that fuch Men were deprived of their natural ftations and rank in life. But in fact it is owing to none of thefe caufes. A fuperior degree of Reafon and Underftanding does not ufually form a Man either for being a more ufeful member of fociety, or more happy in himfelf. Thefe talents are ufually diffipated in fuch a way, as renders them of little account, either to the public or to the poffeffor.—This wafte of Genius exhibits a moft aftonifhing and melancholy profpect. A large library gives a full view of it. Among the multitude of books of which it is compofed, how few engage any one's attention? Such as are addreffed to the heart and imagination, fuch as paint life and manners in juft colours and interefting fituations, and the very few that give genuine defcriptions of Nature

in

in any of her forms, or of the useful and elegant arts, are read and admired. But the far more numerous volumes, productions of the intellectual powers, profound systems and disquisitions of philosophy and theology, are neglected and despised, and remain only as monuments of the pride, ingenuity, and impotency of Human Understanding. Yet many of the inventors of these systems discover the greatest acuteness and depth of Genius; half of which, exerted on any of the useful or elegant arts of life, would have rendered their names immortal. — But it has ever been the misfortune of philosophical Genius to grasp at objects which Providence has placed beyond its reach, and to ascend to general principles and to build systems, without that previous large collection and proper arrangement of facts, which alone can give them a solid foundation. — Notwithstanding this was pointed out by Lord Bacon, in the fullest and clearest manner, yet no attempts have been made to cultivate any one branch of useful philosophy

losophy upon his excellent plan, except by Sir Isaac Newton, Mr. Boyle, and a very few others.—Genius is naturally impatient of restraint, keen and impetuous in its pursuits; it delights therefore in building with materials which the Mind contains within itself, or such as the Imagination can create at pleasure. But the materials requisite for the improvement of any useful art or science, must all be collected from without, by such slow and patient observation, as little suits the vivacity of Genius, and generally requires more bodily activity, than is usually found among Philosophers.

Almost the only pure productions of the Understanding, that have continued to command respect, are those of Abstract Mathematics. These will always be valuable, independent of their application to the useful arts. The exercise they give to the invention, and the agreeable surprise they excite in the Mind, by exhibiting unexpected relations of figures and quantity, are of themselves natural sources of pleasure. This is the only science,

science, the principles of which the philosopher carries in his own Mind; infallible principles to which he can safely trust.

Tho' Men of Genius cannot bear the fetters of method and system, yet they are the only proper people to plan them out. The Genius to lead and direct in philosophy is distinct from, and almost incompatible with the Genius to execute. Lord Bacon was a remarkable instance of this. He brought the Systematic Method of the Schoolmen, which was founded on Metaphysical and often Nominal Subtilties, into deserved contempt, and laid down a method of investigation founded on the justest and most enlarged views of Nature, but which neither himself nor succeeding philosophers have had patience to put in strict execution.

For the reasons above mentioned, it will be found that scarcely any of the useful arts of life owe their improvements to philosophers. They have been principally obliged to accidental discoveries, or to the happy natural sagacity of Men, who exercised cifed thofe arts in private, and who were unacquainted with and undebauched by philofophy.—This has in a particular manner been the fate of Medicine, the moft ufeful of all thofe arts. If by Medicine be meant the art of preferving health, and reftoring it when loft, any Man of fenfe and candor, who has been regularly bred to it, will own that his time has been moftly taken up with enquiries into branches of learning, which upon trial he finds utterly unprofitable to the main ends of his profeffion, or wafted in reading ufelefs theories and voluminous explanations and commentaries on thefe theories; and will ingenuoufly acknowledge, that every thing ufeful, which he ever learned from books in the courfe of many years ftudy, might be taught to any Man of common fenfe and attention in almoft as many months, and that a few years experience is worth all his library.—Medicine in reality owes more to that illiterate enthufiaft, Paracelfus, for introducing fome of the moft ufeful remedies, than to any phyfician who has wrote since the days of Hippocrates, if we except Dr. Sydenham; who owes his reputation entirely to a great natural sagacity in making obfervations, and to a still more uncommon candor in relating them. What little medical philosophy he had, which was as good as his time afforded, served only to warp his Genius, and render his writings more perplexed and tiresome.

But what shews in the strongest light at what an aweful distance philosophers have usually kept from enquiries of general utility to mankind, is, that Agriculture, as a science, is yet only in its infancy.—A mathematician or philosopher, if he happens to possess a farm, does not understand the construction of his cart or plough so well as the fellow who drives them, nor is he so well acquainted with the method of cultivating his ground to the greatest advantage. We have indeed many Systems of Agriculture, that is, we have large compilations of general maxims and principles, along with a profusion of what is called philosophical

phical reafoning on the fubject. But the capital deficiency in Hufbandry is, a copious Collection of particular Obfervations and Experiments, fully and clearly narrated, well attefted, and properly arranged. Thefe alone can give any authority to general Maxims. Without thefe we ought to diftruft all fuch Maxims, as we know many of them are founded on facts, either totally falfe or very imperfectly narrated, and that others are eftablifhed on very erroneous reafoning from facts that are indeed unqueftionable.

It is with pleafure, however, that we obferve the Genius of a more enlarged philofophy arifing, a philofophy fubfervient to life and public utility. Since knowledge has come to be more generally diffufed, that fpirit of free enquiry, which formerly employed itfelf in theology and politics, begins now to pierce into other fciences. The authority of antiquity and great names, in fubjects of opinion, is lefs regarded. Men begin to be weary of theories which lead to no ufeful confequences, and

and have no foundation but in the imagination of ingenious Men. The load of learned rubbiſh, under which ſcience has lain ſo long concealed, partly for the meaneſt and vileſt purpoſes, begins to be taken off; and there ſeems to be a general diſpoſition in Mankind to expoſe to their deſerved contempt thoſe quackiſh and unworthy arts, which have ſo often diſgraced literature and gentlemen of a liberal profeſſion. The true and only method of promoting ſcience, is to communicate it with clearneſs and preciſion, and in a language as much diveſted of technical terms as the nature of the ſubject will admit. What renders this particularly neceſſary is, that ſpeculative Men, who have a Genius for arrangement, and for planning uſeful enquiries, are very often, for reaſons before given, deficient in the executive part. The principles therefore of every ſcience ſhould be explained by them with all poſſible perſpicuity, in order to render them more generally underſtood, and to make their application to the uſeful

arts

arts more eaſy. We have a ſtriking inſtance of the good effects of this, in Chymiſtry. This ſcience lay for many ages involved in the deepeſt obſcurity, concealed under a jargon intelligible to none but a few adepts, and, by a ſtrange aſſociation, frequently interwoven with the wildeſt religious enthuſiaſm. Boerhaave had the very high merit of reſcuing it from this obſcurity, and of explaining it in a language intelligible to every man of common ſenſe. Since that time, Chymiſtry has made very quick advances. The French Philoſophers, in particular, have deſerved well of Mankind for their endeavours to render this ſcience, as well as every branch of natural philoſophy, ſubſervient to the uſeful and elegant arts; and have the additional merit of communicating their knowledge in the eaſieſt and moſt agreeable manner. Mr. Buffon has not only given us the beſt natural hiſtory, but, by the beauty of his compoſition and elegance of his ſtile, has rendered a ſubject, which, in moſt hands, has proved a very dry one, both pleaſing and intereſting.

The ſame liberal and manly ſpirit of enquiry which has diſcovered itſelf in other branches of knowledge, begins to find its way into Medicine. Greater attention is now given to experiment and obſervation; the inſufficiency of an idle theory is more quickly detected, and the pedantry of the profeſſion meets with its deſerved ridicule. We cannot avoid mentioning here, for the honour of our own country, that Pharmacy has been lately reſcued from a ſtate that was a ſcandal to Phyſic and common ſenſe, and is now brought into a judicious, conciſe, and tolerably elegant ſyſtem. Even Agriculture, the moſt natural, the moſt uſeful, and, among the moſt honourable becauſe moſt independent employments, which many years ago began to engage the attention of gentlemen, is now thought a ſubject not unworthy the attention of philoſophers. Mr. du Hamel, who is the Dr. Hales of France, has ſet a noble example in this way, as he does

does in promoting every other branch of knowlege connected with public utility*.

Nothing contributes more to deprive the world of the fruits of great parts, than the paffion for univerfal knowledge, fo conftantly annexed to thofe who poffefs them. By means of this the flame of Genuis is wafted in the endlefs labour of accumulating promifcuous or ufelefs facts, while it might have enlightened the moft ufeful arts by concentrating its force upon a fingle object. This diffipation of Genius is moft effectually checked by the honeft love of fame, which prompts a Man to appear in the world as an author. This neceffarily circumfcribes his excurfions, and determines the force of his Genius to one point. This likewife

---

* His example has been followed by fome others in his own Country and in Switzerland; but in Britain the genuine Spirit of Experimental Agriculture begins to diffufe itfelf with a zeal and rapidity that promifes foon to eftablifh this Science on the moft folid foundation: the public lies under particular obligations, on this fubject, to the fpirit, ingenuity, and induftry of Mr. Young.

likewife refcues him from that ufual abufe and proftitution of fine parts, the wafting of the greateft part of his time in reading, which is really the effect of lazinefs. Here the Mind, being in a great meafure paffive, becomes furfeited with knowledge which it never digefts: the memory is burdened with a load of nonfenfe and impertinence, while the powers of Genius and Invention languifh for want of exercife.

Having obferved of how little confequence a great underftanding generally is to the public, let us next confider the effects it has in promoting the happinefs of the individual.— It is very evident that thofe who devote moft of their time to the exercifes of the Underftanding, are far from being the happieft Men. They enjoy indeed the pleafure arifing from the purfuit and difcovery of Truth. Perhaps too the vanity arifing from a confcioufnefs of fuperior talents adds not a little to their happinefs. But there are many natural fources of pleafure from which they are in a great meafure cut off.— All the public and focial affections, in common with every Tafte natural to the Human Mind, if they are not properly exercifed, grow languid. People who devote moft of their time to the cultivation of their Underftandings, muft of courfe live retired and abftracted from the world. The focial affections (thofe inexhauftible fources of happinefs) have therefore no play, and confequently lofe their natural warmth and vigour. The private and felfifh affections however are not proportionably reduced. Envy and Jealoufy, the moft ungenerous and moft tormenting of all paffions, prevail remarkably among this rank of Men.

Hence perhaps there is lefs friendfhip among learned Men, and efpecially among Authors, than in any other clafs of Mankind. People of independent fortunes, who have no views of intereft or ambition to gratify, naturally connect themfelves with fuch as refemble them in their taftes and fentiments, and as their purfuits do not interfere, their friendfhips may be fincere and lafting.

ing. In thofe profeffions likewife where Intereft is confidered as the immediate object, we often find Men very cordially attached to one anthor, if the field be large enough to admit them all. But in the purfuits of Fame and Vanity, the cafe is very different. There is a jealoufy here that admits no rival, that makes people confider whatever is given to others as taken away from themfelves. Hence the expreffive filence, or the cold, extorted, meafured approbation, given by rival authors to thofe works of Genius, which more impartial and difinterefted Judges receive with the warmeft and moft unreferved applaufe. Such a generofity, fuch a greatnefs of Soul, as render one fuperior to fo mean a jealoufy, are perhaps the rareft Virtues to be found among Mankind.

This ftate of war among Men of Genius and Learning, not only prevents each of them in fome meafure from receiving that portion of Fame to which he is juftly entitled, but is one of the principal caufes which exclude them from that influence and

and afcendency in the different profeffions and affairs of life, which their fuperior talents would otherwife readily procure them. Dull people, though they do not comprehend Men of Genius, are afraid of them, and naturally unite againſt them, and the mutual jealoufies and diffentions among fuch Men, give the dunces all the advantages they could wifh for. As the focial affections become languid, among thofe who devote their whole time to fpeculative fcience, becaufe they are not exercifed, the public affections, the love of liberty and of a native country, become feeble for the fame reafon. There are perhaps no Men who embrace fentiments of patriotifm and public liberty with fo much ardor, as thofe who are juft entering upon the world, and who have got a very liberal and claffical education. Youth indeed is the feafon when every generous and elevated fentiment moft eafily finds its way to the heart: at this happy period, that high fpirit of independence, that zeal for the public, which animated the
Greek

Greek and Roman people, communicate themselves to the soul with a peculiar warmth and enthusiasm. But this fervor too soon subsides. If young men engage in public and active life, every manly and disinterested purpose is in danger of being lost, amidst the universal dissipation and corruption of manners, that surround them; a depravity of manners now become so enormous that any pretension to public Virtue is considered either as hypocrisy or folly. If, on the other hand, they devote themselves to a speculative, sedentary life, abstracted from Society, all the active Virtues and active Powers of the Mind are still more certainly extinguished. A capacity for vigorous and steady exertions can only be preserved by regular habits of Activity. Love of a Country and of a Public cannot subsist among Men, who neither know nor love the individuals which compose that Public. If a man has a family and friends, these give him an interest in the Community, and attach him to it; because their honour and happiness,

which

which he regards as much as his own, are eſſentially connected with its welfare. But if he is a ſingle, ſolitary Being, unconnected with family or friends, there is little to attach him to one country in preference to another. If any encroachment is threatened againſt his perſonal liberty or property, he may think it more eligible to convey himſelf to another country, where he can live unmoleſted, than to ſtruggle, at the riſk of his life and fortune, againſt ſuch encroachments at home. Beſides, we generally find retired ſpeculative Men, who value themſelves on their literary accompliſhments, very much out of humour with the world, if it has not rewarded them according to their own ſenſe of their importance, which it is ſeldom poſſible to do. Swollen with pride and envy, they range all mankind into two claſſes, the Knaves and the Fools. But how can we ſuppoſe one ſhould love a Country or a Community conſiſting of ſuch worthleſs Members?

When abſtraction from company is carried far, it occaſions groſs ignorance.

rance of life and manners, and necessarily deprives a Man of all those little accomplishments and graces which are essential to polished and elegant society, and which can only be acquired by mixing with the world. The want of these is often an insuperable bar to the advancement of persons of real merit, and proves therefore a frequent source of their disgust at the world, and consequently at themselves; for no Man can be happy in himself, who thinks ill of every one around him.

The general complaint of the neglect of merit does not seem to be well founded. It is unreasonable for any Man, who lives detached from society, to complain that his merit is neglected, when he never has made it known. The natural reward of mere Genius, is the esteem of those who know and are judges of it. This reward is never withheld. There is a like unreasonable complaint, that little regard is commonly paid to the good qualities of the heart. But it should be considered, that the world cannot see into the heart,

heart, and can therefore only judge of its goodnefs by vifible effects. There is a natural and proper expreffion of good affections, which ought always to accompany them, and in which true politenefs principally confifts. This expreffion may be counterfeited, and fo may obtain the reward due to genuine virtue; but where this natural index of a worthy character is wanting, or where there is even an outward expreffion of bad difpofitions, the world cannot be blamed for judging from fuch appearances.

Bad health is another common attendant on great parts, when thefe parts are exerted, as is ufually the cafe, rather in a fpeculative than active life.—It is obferved that great quicknefs and vivacity of Genius is commonly attended with a remarkable delicacy of conftitution, and a peculiar fenfibility of the nervous fyftem, and that thofe, who poffefs it, feldom arrive at old age. A fedentary, ftudious life greatly increafes this natural weaknefs of conftitution, and brings on that train of nervous complaints

complaints and low spirits, which render life a burden to the possessor and useless to the public. Nothing can so effectually prevent this as activity, regular exercise, and frequent relaxations of the Mind from those keen pursuits it is usually engaged in.—Too assiduous an exertion of the Mind on any particular subject, not only ruins the health, but impairs the Genius itself; whereas, if the Mind be frequently unbent by amusements, it always returns to its favourite object with double vigour.

But one of the principal misfortunes of a great Understanding, when exerted in a speculative rather than in an active sphere, is its tendency to lead the Mind into too deep a sense of its own weakness and limited capacity. It looks into Nature with too piercing an eye, discovers every where difficulties imperceptible to a common Understanding, and finds its progress stopt by obstacles that appear insurmountable. This naturally produces a gloomy and forlorn Scepticism, which poisons the chearfulness of the temper;

per, and, by the hopeless prospect it gives of improvement, becomes the bane of science and activity. This Sceptical Spirit, when carried into life, renders even Men of the best Understanding unfit for business. When they examine with the greatest accuracy all the possible consequences of a step they are ready to make in life, they discover so many difficulties and chances against them, whichsoever way they turn, that they become slow and fluctuating in their resolutions, and undetermined in their conduct. But as the business of life is in reality only a conjectural art, in which there is no guarding against all possible contingences, a Man that would be useful to the public or to himself, must be at once decisive in his resolutions, and steady and fearless in carrying them into execution.

We shall mention, in the last place, among the inconveniences attendant on superior parts, that solitude in which they place a person on whom they are bestowed, even in the midst of society.

Condemned

Condemned in Bufinefs or in Arts to drudge,
Without a Second and without a Judge *.

To the few, who are judges of his abilities, he is an object of jealoufy and envy. The bulk of Mankind confider him with that awe and diftant regard that is incompatible with confidence and friendfhip. They will never unbofom themfelves to one they are afraid of, nor lay open their weakneffes to one they think has none of his own. For this reafon we commonly find that even Men of Genius have the greateft real affection and friendfhip for fuch as are very much their inferiors in point of Underftanding; good-natured, unobferving people, with whom they can indulge all their peculiarities and weakneffes without referve. Men of great abilities therefore, who prefer the fweets of focial life and private friendfhip to the vanity of being admired, ought carefully to conceal their fuperiority, and bring themfelves down to the level of thofe they converfe with. Nor muft this feem to be the effect of a defigned

* Pope.

designed condescension; for that is peculiarly mortifying to human pride.

Thus we have endeavoured to point out the effects which the faculty of Reason, that boasted characteristic and privilege of the Human Species, produces among those who possess it in the most eminent degree: and, from the little influence it seems to have in promoting either public or private good, we are almost tempted to suspect, that Providence deprives us of those fruits we naturally expect from it, in order to preserve a certain ballance and equality among Mankind.—Certain it is that Virtue, Genius, Beauty, Wealth, Power, and every natural advantage one can be possessed of, are usually mixed with some alloy, which disappoints the fond hope of their raising the possessor to any uncommon degree of eminence, and even in some measure brings him down to the common level of his Species.

The next distinguishing principle of Mankind, which was mentioned, is that which unites them into societies, and attaches them to one another

ther by sympathy and affection. This principle is the source of the most heart-felt pleasure which we ever taste.

It does not appear to have any natural connection with the Understanding.—It was before observed that persons of the best Understanding possessed it frequently in a very inferior degree to the rest of Mankind; but it was at the same time mentioned that this did not proceed from less natural sensibility of heart, but from the Social Principle languishing for want of proper exercise. By its being more exercised among the idle and the dissipated, persons of this character sometimes derive more pleasure from it; for not only their pleasures but their vices are often of the social kind; and hence the Social Principle is warm and vigorous among them. Even drinking, if not carried to excess, is found favourable to this principle, especially in our northern climates, where the affections are naturally cold; as it produces an artificial warmth of temper, opens and enlarges the heart,

and

and difpels the referve, natural perhaps to wife Men, but inconfiftent with connections of fympathy and affection.

All thofe warm and elevated defcriptions of friendfhips, which fo powerfully charm the Minds of young people, and reprefent it as the height of human felicity, are really romantic among us. When we look round us into life, we meet with nothing correfponding to them, except among an happy few in the fequeftered fcenes of life, far removed from the purfuits of intereft or ambition. Thefe fentiments of friendfhip are original and genuine productions of warmer and happier climes, and adopted by us merely out of vanity.—The fame obfervation may be applied to the more delicate and interefting attachment between the fexes.—Many of our fex, who, becaufe poffeffed of fome learning, affume the tone of fuperior wifdom, treat this attachment with great ridicule, as a weaknefs below the dignity of a Man, and allow no kind of it but what we have in common

mon with the whole Animal Creation. They acknowledge, that the fair sex are useful to us, and a very few will deign to consider some of them as reasonable and agreeable companions.—But it may be questioned, whether this is not the language of an heart insensible to the most refined and exquisite pleasure Human Nature is capable of enjoying, or the language of disappointed Pride, rather than of Wisdom and Nature. No Man ever despised the sex who was a favourite with them, nor did any one ever speak contemptuously of love, who was conscious of loving and being beloved by a Woman of merit. The attachment between the sexes is a natural principle, which forms in an eminent degree the happiness of Human Life in every part of the world. As the power of beauty in the Eastern countries is extremely absolute, no other accomplishments are thought necessary to the women, but such as are merely personal. They are cut off therefore, by the most cruel exertion of power, from all opportunities

nities of improvement, and pafs their lives in a lonely and ignominious confinement; excluded from all free intercourfe with human fociety. The cafe is very different in this climate, where the power of Beauty is very limited. Love with us is but a feeble paffion, and generally yields eafily to intereft, ambition, or even to vanity, that paffion of a little mind and a cold heart; as luxury therefore advances among us, love muft be extinguifhed among people of better rank altogether. To give it any force or permanency, we muft connect it with fentiment and efteem. But it is not in our power to do this, if we treat Women as we do Children. If we imprefs their minds with a belief that they were only made to be domeftic drudges, and the flaves of our pleafures, we debafe their minds, and deftroy all generous emulation to excel; whereas, if we ufe them in a more liberal and generous manner; a decent pride, a confcious dignity, and a fenfe of their own worth, will naturally induce them to exert themfelves

to be what they would wish to be thought, and are entitled to be, our companions and friends. This however they can never accomplish by leaving their own natural characters and assuming ours. As the two sexes have very different parts to act in life, Nature has marked their characters very differently; in a way that best qualifies them to fulfil their respective duties in society. Nature intended us to protect the Women, to provide for them and their families. Our business is without doors. All the rougher and more laborious parts in the great scene of human affairs fall to our share. In the course of these, we have occasion for our greater bodily strength, greater personal courage, and more enlarged powers of Understanding. The greatest glory of Women lies in private and domestic life, as friends, wives, and mothers. It belongs to them, to regulate the whole œconomy of the family. But a much more important charge is committed to them. The education of the youth of both sexes principally devolves
upon

upon the Women, not only in their infancy, but during that period, in which the conſtitution both of body and mind, the temper and difpofitions of the heart, are in a great meaſure formed. They are defigned to ſoften our hearts and poliſh our manners. The form of power and authority, to direct the affairs of public focieties and private families, remains indeed with us. But they have a natural defence againſt the abuſe of this power, by that ſoft and infinuating addreſs, which enables them to controul it, and often to transfer it to themſelves.

In this view, the part which women have to act in life, is important and refpectable; and Nature has given them all the neceſſary requifites to perform it. They poſſeſs, in a degree greatly beyond us, fenfibility of heart, ſweetneſs of temper, and gentleneſs of manners. They are more chearful and joyous. They have a quicker difcernment of characters. They have a more lively fancy, and a greater delicacy of taſte and ſentiment; they are better judges

judges of grace, elegance, and propriety, and therefore are our superiors in such works of taste as depend on these. If we do not consider Women in this honourable point of view, we must forego in a great measure the pleasure arising from an intercourse between the sexes, and, together with this, the joys and endearments of domestic life. Besides, in point of sound policy, we should either improve the Women or abridge their power; if we give them an important trust, we should qualify them for the proper discharge of it; if we give them liberty, we should guard against their abuse of it; and not trust so entirely as many of us do to their insensibility or to their religion. A Woman of a generous spirit, if she is treated as a friend and an equal, will feel and gratefully return the obligation; and a Man of a noble mind will be infinitely more gratified with the attachment of a Woman of merit, than with the obedience of a dependant and a slave.

If we enquire into the other pleafures we enjoy as Social Beings, we fhall find many delicacies and refinements admired by fome, which others who never felt them, treat as vifionary and romantic. It is no difficult matter to account for this. There is certainly an original difference in the conftitutions both of Men and of Nations; but this is not fo great as at firft view it feems to be. Human Nature confifts of the fame principles every where. In fome people one principle is naturally ftronger than it is in others, but exercife and proper culture will do much to fupply the deficiency. The inhabitants of cold climates, having lefs natural warmth and fenfibility of heart, enter but very faintly into thofe refinements of the Social Principle, in which Men of a different temper delight. But if fuch refinements are capable of affording to the Mind innocent and fubftantial pleafure, it fhould be the bufinefs of philofophy to fearch into the proper methods of cultivating and improving them. This ftudy, which makes a confiderable

able part of the philosophy of life and manners, has been surprisingly neglected in Great Britain. Whence is it that the English, with great natural Genius and Acuteness, and still greater Goodness of heart, blessed with riches and liberty, are rather a melancholy and unhappy people? Why is their neighbouring nation, whom they despise for their shallowness and levity, yet awkwardly imitate in their most frivolous accomplishments, happy in poverty and slavery? We are obliged to own the one possesses a native chearfulness and vivacity, beyond any other people upon earth; but still much is owing to their cultivating with the greatest care all the arts which enliven and captivate the imagination, soften the heart, and give society its highest polish. In Britain we generally find Men of sense and learning speaking in a contemptuous manner of all writings addressed to the imagination and the heart, even of such as exhibit genuine pictures of life and manners. But besides the additional vigour, which these give
to

to the powers of the imagination, and the influence they have in rendering the affections warmer and more lively, they are frequently of the greateſt ſervice in communicating a knowledge of the world: a knowledge the moſt important of all others, to one who is to live in it, and who would wiſh to act his part with propriety and dignity. Moral painting is undoubtedly the higheſt and moſt uſeful ſpecies of painting. The execution may be, and generally is, very wretched, and ſuch as has the worſt effects, in miſleading the judgment and debauching the heart: but, if this kind of writing continues to come into the hands of Men of Genius and worth, little room will be left for this complaint.

There is a remarkable difference between the Engliſh and French in their taſte of ſocial life. The gentlemen in France, in all periods of life, and even in the moſt advanced age, never aſſociate with one another, but ſpend all the hours they can ſpare from buſineſs or ſtudy with the ladies; with the young, the gay, and the happy.—It is observed that the people of this rank in France live longer, and, what is of much greater consequence, live more happily, and enjoy their faculties of body and Mind more entire, in old age, than any people in Europe. In Great Britain we have certain notions of propriety and decorum, which lead us to think the French manner of spending their hours of relaxation from business extremely ridiculous. But if we examine with due attention into these sentiments of propriety, we shall not perhaps find them to be built on a very solid foundation. We believe that it is proper for persons of the same age, of the same sex, of similar dispositions and pursuits, to associate together. But here we seem to be deceived by words. If we consult nature and common sense, we shall find that the true propriety and harmony of social life consists in the association of people of different dispositions and characters, judiciously blended together. Nature has made no individual, nor any class of people, independent of the rest

rest of their species, or sufficient for their own happiness. Each sex, each character, each period of life, have their several advantages and disadvantages; and that union is the happiest and most proper, where wants are mutually supplied. The fair sex should naturally expect to gain, from our conversation, knowledge, wisdom, and sedateness; and they should give us in exchange humanity, politeness, chearfulness, taste, and sentiment. The levity, the rashness, and the folly of early life, is tempered with the gravity, the caution, and the wisdom of age; while the timidity, coldness of heart, and languor, incident to declining years, are supported and assisted by the courage, the warmth, and the vivacity of youth.

Old people would find great advantage in associating rather with the young than with those of their own age.—Many causes contribute to destroy chearfulness in the decline of life, besides the natural decay of youthful vivacity. The few surviving friends and companions are then dropping

dropping off apace; the gay prospects, that swelled the imagination in more early and more happy days, are then vanished, and, together with them, the open, generous, unsuspicious temper, and that warm heart which dilated with benevolence to all mankind. These are succeeded by gloom, disgust, suspicion, and all the selfish passions which sour the temper and contract the heart. When old people associate only with one another, they mutually increase these unhappy dispositions, by brooding over their disappointments, the degeneracy of the times, and such like chearless and uncomfortable subjects. The conversation of young people dispels this gloom, and communicates a chearfulness, and something else perhaps which we do not fully understand, of great consequence to health and the prolongation of life. There is an universal principle of imitation among Mankind, which disposes them to catch instantaneously, and without being conscious of it, the resemblance of any action or character that presents fents itself. This difpofition we can often check by the force of Reafon, or the affiftance of oppofite impreffions: at other times, it is infurmountable. We have numberlefs examples of this in the fimilitude of character and manners induced by people living much together, in the fudden communications of terror, of melancholy, of joy, of the military ardor, when no caufe can be affigned for thefe emotions. The communication of nervous diforders, efpecially of the convulfive kind, is often fo aftonifhing, that it has been referred to fafcination or witchcraft. We fhall not pretend to explain the nature of this mental infection; but it is a fact well eftablifhed, that fuch a thing exifts, and that there is fuch a principle in nature as an healthy fympathy, as well as a morbid infection.

An old Man, who enters into this philofophy, is far from envying or proving a check on the innocent pleafures of young people, and particularly of his own Children. On the contrary, he attends with delight to the gradual opening of the Imagination and the dawn of Reason; he enters by a secret sort of sympathy into their guiltless joys, that recall to his memory the tender images of his youth, which, by length of time, have contracted a * softness inexpressibly agreeable; and thus the evening of life is protracted to an happy, honourable, and unenvied old age.

* Addison.

SECTION III.

THE advantages derived to Mankind from Taste, by which we understand the improved use of the powers of the imagination, are confined to a very small number. Taste implies not only a quickness and justness of intellectual discernment, but also a delicacy of feeling in regard to pleasure or pain, consequent upon a discernment of its proper object. The servile condition of the bulk of Mankind requires constant labour for their daily subsistence. This of necessity deprives them of the means of improving the powers either of Imagination or of Reason, except so far as their particular employments render such an improvement necessary. Yet there is great reason to think the Men of this class the happiest, at least such of them as are just above

above want. If they do not enjoy the pleasure arising from the proper culture of the higher powers of their Nature, they are free from the misery consequent upon the abuse of these powers. They are likewise in full possession of one great source of human happiness: which is good health and good spirits. Their Minds never languish for want of exercise or want of a pursuit, and therefore the tædium vitæ, the insupportable listlessness arising from the want of something to wish or something to fear, is to them unknown.

But even among those to whom an easy fortune gives sufficient leisure and opportunities for the improvement of Taste, we find little attention given to it, and consequently little pleasure derived from it. Nature gives only the seeds of Taste, culture must rear them, or they will never become a considerable source of pleasure. The only powers of the Mind, that have been much cultivated in this Island, are those of the Understanding. One unhappy consequence of this has been

been to diffolve the natural union between philofophy and the fine arts; an union extremely neceffary to their improvement. Hence Mufic, Painting, Sculpture, Architecture, have been left in the hands of ignorant artifts unaffifted by philofophy, and even unacquainted with the works of great mafters.

The productions of purely natural Genius are fometimes great and furprifing, but are generally attended with a wildnefs and luxuriancy inconfiftent with juft Tafte. It is the bufinefs of philofophy to analyfe and afcertain the principles of every art where Tafte is concerned; but this does not require a philofopher to be mafter of the executive part of thefe arts, or to be an inventor in them. His bufinefs is to direct the exertion of Genius in fuch a manner that its productions may attain to the utmoft poffible perfection.

It is but lately that any attempt was made among us to analyfe the principles of beauty, or of mufical expreffion. And its having been made was entirely owing to the accident

cident of two eminent artiſts, the one in Painting*, the other in Muſic †, having a philoſophical ſpirit, and applying it to their ſeveral profeſſions. Their being eminent maſters and performers, was undoubtedly of ſingular advantage to them in writing on theſe ſubjects, but was by no means ſo eſſential as is generally believed. Mr. Webb, who was no painter, has explained the principles of Taſte in painting with an accuracy and perſpicuity, which would have done honour to the greateſt maſter. He ſhews at the ſame time, that if we are wholly guided by the prejudice of names, we no longer truſt our own ſenſes; that we muſt acknowledge merit which we do not ſee, and undervalue that which we do; and that diſtreſſed between authority and conviction, we become diſguſted with the difficulty of an art, which is perhaps of all others the moſt eaſily underſtood, becauſe it is the moſt direct and immediate addreſs to the ſenſes.

It

* Hogarth.   † Aviſon.

It is likewife but very lately that modern philofophy has condefcended to beftow any attention on poetry or compofition of any kind. The genuine fpirit of criticifm is but juft beginning to exert itfelf. The confequence has been, that all thefe arts have been under the abfolute dominion of fafhion and caprice, and therefore have not given that high and lafting pleafure to the Mind, which they would have done, if they had been exercifed in a way agreeable to Nature and juft Tafte.

Thus in painting, the fubject is very feldom fuch as has any grateful influence on the Mind. The defign and execution, as far as the mere painter is concerned, is often admirable, and the tafte of imitation is highly gratified, but the whole piece wants meaning and expreffion, or what it has is trifling and often extremely difagreeable. It is but feldom we fee nature painted in her moft amiable or graceful forms, in a way that may captivate the heart and make it better. On the contrary, we often find her in fituations the moft most unpleasing to the Mind, in old-age, deformity, disease, and idiotism. The Dutch and many of the Flemish commonly exhibit her in the lowest and most debasing attitudes; and in Italy the Genius of painting is frequently prostituted to the purposes of the most despicable superstition.— Thus the Mind is disappointed in the pleasure which this elegant art is so admirably fitted to convey; the agreeable effect of the imitation being counteracted and destroyed by the unhappy choice of the subject.

The influence of Music over the Mind is perhaps greater than that of any of the fine arts. It is capable of raising and soothing every passion and emotion of the soul. Yet the real effects produced by it are inconsiderable. This is in a great measure owing to its being left in the hands of practical Musicians, and not under the direction of Taste and Philosophy: For, in order to give Music any extensive influence over the Mind, the composer and

performer

performer muſt underſtand well the human heart, the various aſſociations of the paſſions, and the natural tranſitions from one to another, ſo as they may be able to command them, in conſequence of their ſkill in muſical expreſſion.

No Science ever flouriſhed, while it was confined to a ſet of Men who lived by it as a profeſſion. Such Men have purſuits very different from the end and deſign of their art. The intereſted views of a trade are widely different from the enlarged and liberal proſpects of Genius and Science. When the knowledge of an art is confined in this manner, every private practitioner muſt attend to the general principles of his craft, or ſtarve. If he goes out of the common path, he is in danger of becoming an object of the jealouſy and the abuſe of his brethren; and among the reſt of Mankind he can neither find judges nor patrons. This is particularly the caſe of the delightful art we are ſpeaking of, which has now become a Science ſcarcely underſtood by any but a

few

few compofers and performers. They alone direct the public Tafte, or rather dictate to the world what they fhould admire and be moved with; and the vanity of moft people makes them acquiefce in this affumed authority, left otherwife they fhould be fufpected to want Tafte and knowledge in the fubject. In the mean time, Men of fenfe and candor, not finding that pleafure in Mufic which they were made to expect, are above diffembling, and give up all pretenfions to the leaft knowledge in the Subject. They are even modeft enough to afcribe their infenfibility of the charms of Mufic to their want of a good ear, or a natural Tafte for it, and own that they find the Science fo complicated, that they do not think it worth the trouble it muft coft them to acquire an artificial one. They refolve to abandon an Art in which they defpair of ever becoming fuch proficients, as either to derive pleafure from it themfelves, or to be able to communicate it to others, at leaft without making that the ferious bufinefs of Life, which
ought

ought only to be the amufement of an idle or the folace of a melancholy hour. But before they entirely forego one of the moft innocent amufements in life, not to fpeak of it in an higher ftile, it would not be improper to enquire a little more particularly into the fubject. We fhall therefore here beg leave to examine fome of the firft principles of Tafte in Mufic with the utmoft freedom.

Mufic is the Science of founds, fo far as they affect the Mind. Nature independent of cuftom has connected certain founds or tones with certain feelings of the Mind. Meafure and proportion in founds have likewife their foundation in Nature. Thus certain tones are naturally adapted to folemn, plaintive, and mournful fubjects, and the movement is flow; others are expreffive of the joyous and elevating, and the movement is quick.— Sounds likewife affect the Mind, as they are loud or foft, rough or fmooth, diftinct from the confideration of their gravity or acutenefs. Thus in the Æolian harp the tones are pleafant and foothing,

though

though there is no fucceffion of notes varying in acutenefs, but only in loudnefs. The effect of the common drum, in roufing and elevating the Mind, is very ftrong; yet it has no variety of notes; though the effect indeed here depends much on the proportion and meafure of the notes.

Melody confifts in the agreeable fucceffion of fingle founds. — The melody that pleafes in one country does not equally pleafe in another, though there are certain general principles which univerfally regulate it, the fcale of Mufic being the fame in all countries. — Harmony confifts in the agreeable effect of founds differing in acutenefs produced together; the general principles of it are likewife fixed.

One end of Mufic is merely to communicate pleafure, by giving a flight and tranfient gratification to the Ear; but the far nobler and more important is to command the paffions and move the heart. In the firft view it is an innocent amufement, well fitted to give an agreeable

ble relaxation to the Mind from the fatigue of ſtudy or buſineſs.—In the other it is one of the moſt uſeful arts in life.

Muſic has always been an art of more real importance among uncultivated than among civilized nations. Among the former we always find it intimately connected with poetry and dancing, and it appears, by the teſtimony of many ancient * authors, that Muſic, in the original ſenſe of the word, comprehended melody, dance and ſong. By theſe almoſt all barbarous nations in every age, and in every climate, have expreſſed all ſtrong emotions of the Mind. By † theſe attractive and powerful arts they celebrate their public ſolemnities; by theſe they lament their private and public calamities, the death of friends or the loſs of warriors; by theſe united they expreſs their joy on their marriages, harveſts, huntings, victories; praiſe the great actions of their gods and heroes; excite each other to war and brave exploits,

* See Plato and Athenæus.   † Brown.

ploits, or to suffer death and torments with unshaken constancy..

In the earliest periods of the Greek states, their most ancient maxims, exhortations, and laws, and even their history, were written in verse, their religious rites were accompanied by dance and song, and their earliest oracles were delivered in verse, and sung by the priest or priestess of the supposed God. While melody, therefore, conjoined with poetry, continued to be the established vehicle of all the leading principles of religion, morals, and polity, they became the natural and proper objects of public attention and regard, and bore a principal and essential part in the * education of Children. Hence we see how Music among the ancient Greeks was esteemed a necessary accomplishment, and why an ignorance in this art was regarded as a capital defect. Thus Themistocles came to be reproached with his ignorance in † Music, and the many enormous crimes committed in the country of Cynethe were attributed

---

* Plutarchus de Musica.   † Cicero.

tributed by the neigbouring ſtates to the neglect of * Muſic; nor was the reproach thrown, in theſe days, upon ſuch as were ignorant of the art, without a juſt foundation; becauſe this ignorance implied a general deficiency in the three great articles of education; religion, morals, and polity.

† Such was the enlarged Nature of ancient Muſic when applied to education, and not a mere proficiency in the playing or ſinging art, as has been very generally ſuppoſed. Moſt authors have been led into this miſtake by Ariſtotle, who ſpeaks of Muſic as an art diſtinct from Poetry. But the reaſon of this was, that in the time of Ariſtotle, a ſeparation of the melody and ſong had taken place; the firſt retained the name of Muſic, and the ſecond aſſumed that of poetry.

In the moſt ancient times the character of a bard was of great dignity and importance, being uſually united with that of legiſlator and chief magiſtrate.

* Athenæus, Polybius.   † See Plato de Legibus.

magiſtrate. Even after the ſeparation was firſt made, he continued for ſome time to be the ſecond character in the community; as an aſſiſtant to the magiſtrate in governing the people. *

Such was the important and honourable ſtate of Muſic, not only in ancient Greece, but in the early periods of all civilized nations in every part of the world.

In all the Celtic nations, and particularly in Great Britain, the bards were anciently of the higheſt rank and eſtimation. The character of general, poet, and muſician, were united in Fingal and † Oſſian. The progreſs of Edward the firſt's arms was ſo much retarded by the influence of the Welſh bards, whoſe ſongs breathed the high ſpirit of liberty and

* Suidas on the Leſbian Song. Heſiod.
† Such was the ſong of Fingal, in the day of his joy. His thouſand bards leaned forward from their ſeats, to hear the voice of the king. It was like the Muſic of the harp on the gale of the ſpring. Lovely were thy thoughts, O Fingal! why had not Oſſian the ſtrength of thy ſoul? but thou ſtandeſt alone, my father; and who can equal the king of Morven? Carthon.

and war, that he bafely ordered them to be flain: an event that has given rife to one of the moſt elegant and fublime odes that any language has produced.

In proportion as the fimplicity and purity of ancient manners declined in Greece, thefe fifter arts, which formerly ufed to be the handmaids of virtue, came by degrees to be proftituted to the purpofes of vice or of mere amufement. A corruption of manners debafed thefe arts, which, when once corrupted, become principal inftruments in compleating the deftruction of religion and virtue. Yet the fame caufe which turned them afide from their original ufe, contributed to their improvement as particular arts. When Mufic, Dancing, and poerty came to be confidered as only fubfervient to pleafure, a higher degree of proficiency in them became neceffary, and confequently a more fevere application to each. This compleated their feparation from one another, and occafioned their falling entirely into the hands of fuch Men as devoted

voted their whole time to their cultivation. Thus the complex character of legiflator, poet, actor and mufician, which formerly fubfifted in one perfon, came to be feparated into diftinct profeffions, and the unworthy purpofes to which Mufic in particular came to be applied, made a * proficiency in it unfuitable to any Man of high rank and character.

Doctor Brown has treated this fubject at full length, in a very learned differtation, where he has fhewn with great ingenuity and by the cleareft deduction from facts, how melody, dance, and fong, came, in the progrefs of civilized fociety, in different nations, to be cultivated feparately; and by what means, upon their total feparation, the power, the utility, and dignity of Mufic, has funk into a general corruption and contempt.

The effect of eloquence depends in a great meafure on Mufic. We take Mufic here in the large and proper fenfe of the word; the art of varioufly

* Ariftot. Politic. Plutar. de Mufica.

ly affecting the Mind by the power of sounds. In this sense, all Mankind are more or less judges of it, without regard to exactness of ear. Every Man feels the difference between a sweet and melodious voice and a harsh dissonant one.

Every agreeable speaker, independent of the sweetness of his tones, rises and falls in his voice in strict musical intervals, and therefore his discourse is as capable of being set in musical characters as any song whatever. But however musical a voice may be, if the intervals which it uses are uniformly the same, it displeases, because the ear is fatigued with the constant return of the same sounds, however agreeable; and if we attend to the subject, we are displeased on another account, at hearing the same musical passages made use of to express and inspire sentiments of the most different and opposite natures; whereas the one should be always varying and adapted to the other. This has justly brought great ridicule on what is called Singing a Discourse,

course, though what really offends is either the badness of the song, or its being tiresome for want of variety.

If we examine into the effects produced by eloquence in all ages, we must ascribe them in a great degree to the power of sounds. We allow, at the same time, that composition, action, the expression of the countenance, and some other circumstances, contribute their share, though a much smaller one.—The most pathetic composition may be pronounced in such a manner, as to prevent its having the least influence. Orations which have commanded the Minds of the greatest Men, and determined the fate of nations, have been read in the closet with languor and disgust.

As the proper application of the voice to the purposes of eloquence has been little attended to, it has been thought an art unattainable by any rules, and depending entirely on natural Taste and Genius. This is in some measure true; yet it is much more reducible to rules, and more
capable

capable of being taught, than is commonly imagined. Indeed, before philofophy afcertains and methodizes the ideas and principles on which an art depends, it is no wonder it be difficult of acquifition. The very language in which it is to be communicated is to be formed, and it is a confiderable time before this language comes to be underftood and adopted.—We have a remarkable inftance of this in the fubject of mufical expreffion, or performing a piece of Mufic with Tafte and propriety. People were fenfible, that the fame Mufic performed by different artifts had very different effects. Yet they all played the fame notes, and played equally well in tune and in time. But ftill there was an unknown fomewhat, that gave it meaning and expreffion from one hand, while from another it was lifelefs and infipid. People were fatisfied in refolving this into performing with or without Tafte, which was thought the entire gift of Nature.—Geminiani, who was both a compofer and performer of the highest class, first thought of reducing the art of playing on the Violin with Taste to rules, for which purpose he was obliged to make a great addition to the musical language and characters. The scheme was executed with great ingenuity, but has not met with the attention it deserved.

Music, like Eloquence, must propose as its end a certain effect to be produced on the hearers. If it produces this effect, it is good Music; if it fails, it is bad. — No Music can be pronounced good or bad in itself; it can only be relatively so. Every country has a melody peculiar to itself, expressive of the several passions. A composer must have a particular regard to this, if he proposes to affect them. — Thus in Scotland there is a chearful Music perfectly well fitted to inspire that joyous mirth suited to dancing, and a plaintive Music peculiarly expressive of that tenderness and pleasing melancholy attendant on distress in love; both original in their kind, and different

ferent from every other in Europe. *
It is of no confequence whence
this

---

* There is a fimplicity, a delicacy, and pathetic expreffion in the Scotch airs, which have always made them admired by people of genuine Tafte in Mufic. It is a general opinion, that many of them were compofed by David Rizzio: but this appears very improbable. There is a peculiarity in the ftile of the Scotch melody, which foreigners, even fome of great knowledge in Mufic, who refided long in Scotland, have often attempted to imitate, but never with fuccefs. It is not therefore probable, that a ftranger, in the decline of life, who refided only three or four years in Scotland, fhould enter fo perfectly into the Tafte of the national Mufic, as to compofe airs, which the nicefl judges cannot diftinguifh from thofe which are certainly known to be of much greater antiquity than Rizzio's. The tradition on this fubject is very vague, and there is no fhadow of authority to afcribe any one particular Scotch air to Rizzio. If he had compofed any Mufic while he was in Scotland, it is highly probable it would have partaken of the genius of that melody, to which he had been accuftomed; but the ftyle of the Scotch and Italian airs, in Rizzio's time, bear not the leaft refemblance to one another. Perhaps he might have moulded fome of the Scotch airs into a more regular form; but if he did, it was probably no real improvement; as the wildeft of them, which bid defiance to all rules of modern compofition, are generally the moft powerfully affecting.

this Music derives its origin, whether it be simple or complex, agreeable to the rules of regular composition, or against them; whilst it produces its intended effect, in a superior degree to any other, it is the preferable Music; and while a person feels this effect, it is a reflection on his Taste and common sense, if not on his candor, to despise it. The Scotch will in all probability soon lose this native Music, the source of so much pleasure to their ancestors, without acquiring any other in its place. Most musical people in Scotland either neglect it altogether, or destroy that simplicity in its performance on which its effects so entirely depended, by a fantastical and absurd addition of Graces foreign to the genius of its Melody. The contempt shewn for the Scotch Music in its primitive and pathetic simplicity, by those who from a superior skill in the science, are thought entitled to lead the public Taste, has nearly brought it into universal discredit. Such is the tyranny of Fashion, and such are the effects of that

that vanity, which determines us, in obedience to its dictates, to refign any pleafure, and to fubmit to almoft any pain.

They who apply much of their time to Mufic, acquire new Taftes, befides their national one, and, in the infinite variety which melody and harmony are capable of, difcover new fources of pleafure formerly unknown to them. But the fineft natural Tafte never adopts a new one, till the ear has been long accuftomed to it; and, after all feldom enters into it with that warmth and feeling, which thofe do to whom it is national.

The general admiration pretended to be given to foreign Mufic in Britain, is in general defpicable affectation. In Italy we fometimes fee the natives tranfported, at the opera, with all that variety of delight and paffion which the compofer intended to produce. The fame opera in England is feen with the moft remarkable liftleffnefs and inattention. It can raife no paffion in the audience, becaufe they do not underftand the language

language in which it is written. To them it has as little meaning as a piece of inſtrumental Muſic. The ear may be tranſiently pleaſed with the air of a ſong; but that is the moſt trifling effect of Muſic. Among the very few who underſtand the language, and enter with pleaſure and taſte into the Italian Muſic, the conduct of the dramatic part appears ſo ridiculous, that they can feel nothing of that tranſport of paſſion, the united effect of Muſic and Poetry, which may be gradually raiſed by the artful texture and unfolding of a dramatic ſtory.* Yet vanity prevails ſo much over the ſenſe of pleaſure itſelf, that the Italian opera is in England more frequented by people of rank, than any other public diverſion; and, to avoid the imputation of want of Taſte, they condemn themſelves to ſome hours painful attendance on it every week, and pretend to talk of it in raptures, to which their hearts will ever remain ſtrangers.

Nothing

* Brown.

Nothing can afford so convincing a proof of the absolute incapacity of our modern Music, to produce any lasting effect on the passions of Mankind, as the observation of the effects produced by an opera on people of the greatest knowledge and Taste in Music, as well as on those who are most ignorant of the science. An affecting story may be wrought up, by the genius of a Metastasio, in a manner that shall make it be read with the highest delight and emotion by every person of Taste and Sensibility. We should naturally suppose that the addition of Music ought to communicate greater energy to the composition; but, instead of this, it totally annihilates it. Many people may return home from an opera with their ears highly gratified by some particular songs, or passages of songs; but never one returned affected with the catastrophe of the piece, or with the heart-felt emotion produced by Othello or King Lear.

Simplicity in melody is absolutely necessary in all Music intended to reach the heart, or even greatly to delight

delight the ear. The effect here must be produced inftantaneoufly, or not at all. The fubject of the Mufic muft therefore be fimple, and eafily traced, and not a fingle note or grace fhould be admitted, but what has a tendency to the propofed end. — If fimplicity of melody be fo neceffary, where the intention is to move the paffions, fimplicity of harmony, which ought always to be fubfervient to it, muft be ftill more neceffary. Some of the moft delicate touches of pathetic Mufic will not allow any accompanyment.

The ancient Mufic certainly produced much greater and more general effects than the modern, though we fhould allow the accounts we have of it to be much exaggerated. Yet the fcience of Mufic was in a very low ftate among the ancients. They were probably ftrangers to harmony, at leaft if they knew it they neglected it, all the voices and inftruments being unifons in concert: and the inftruments they made ufe of, appear to have been much inferior, in refpect of compafs, expreffion,

preffion, and variety, to thofe which we are poffeffed of. Yet thefe very deficiencies might render their Mufic more expreffive and powerful. The only view of compofers was to touch the heart and the paffions. Simple melody was fufficient for this purpofe, which might eafily be comprehended and felt by the whole people. There were not two different fpecies of Mufic among them, as with us, one for the learned in the fcience, and another for the vulgar.

* Although we are ignorant of the particular conftruction of the ancient Mufic, yet we know it muft have been altogether fimple; fuch as ftatefmen, warriors, and bards, occupied in other purfuits, could compofe, and fuch as people of all ranks, children, and men bufied in other concerns of life, could learn and practice. We are likewife ftrangers to the particular ftructure of their inftruments, but we have the greateft reafon to believe they were extremely fimple. The chords of the lyre were

* Brown.

were originally but four *. They were afterwards increased to seven, at which number they were fixed by the laws of Sparta †, and Timotheus was banished for adding four additional strings; but we are uncertain of the intervals by which the strings of the lyre ascended. Those who regard only the advancement of Music as a science, treat the laws of Sparta upon this subject with great ridicule; but they who consider it as an art intimately connected with the whole fabric of its religion, Morals, and policy, will view them in a very different light, and see the necessity of preserving their Music in the utmost degree of simplicity. In fact, when the lyre, in process of time, acquired forty strings, when Music came to be a complicated art, and to be separately cultivated by those who gave up their whole time to its improvement, its noblest end and aim was lost. In ‡ Plutarch's time

* Pausanias.

† The art of Music had formerly been fixed and made unalterable in Crete and Egypt. Plato de legibus.

‡ De Musica.

time it was funk into a mere amufement of the theatre. The fame caufes have produced the fame effects in modern times. In proportion as Mufic has become more artificial, and more difficult in the execution, it has loft of its power and influence.

It was formerly obferved, that the power of the ancient melody depended much on its union with poetry. There are other circumftances which might contribute to this power. The different paffions naturally exprefs themfelves by different founds; but this expreffion feems capable of a confiderable latitude, and may be much altered by early affociation and habit. When particular founds and a certain ftrain of melody are impreffed upon young minds, in a uniform connexion with certain paffions expreffed in a fong, this regular affociation raifes thefe founds, in progrefs of time, into a kind of natural and expreffive language of the paffions. * Melody therefore is to be confidered, in a certain degree, as a relative thing, founded in the

* Brown.

the particular affociation and habits of different people; and, by cuftom, like language, annexed to their fentiments and paffions. We generally hear with pleafure the Mufic we have been accuftomed to in our youth, becaufe it awakes the memory of our guiltlefs and happy days. We are even fometimes wonderfully affected with airs, that neither appear, to ourfelves nor to others, to have any peculiar expreffion. The reafon is, we have firft heard thefe airs at a time when our minds were fo deeply affected by fome paffions, as to give a tincture to every object that prefented itfelf at the fame time; and though the paffion and the caufe of it are entirely forgot, yet an object that has once been connected with them, will often awake the emotion, though it cannot recall to remembrance the original caufe of it.

* Similar affociations are formed, by the appropriations, in a great meafure accidental, which different nations have given to particular mufical inftruments, as bells, drums, trumpets,

---

* Brown.

trumpets, and organs; in confequence of which they excite ideas and paffions in fome people which they do not in others. No Englifhman can annex warlike ideas to the found of a bagpipe.

We have endeavoured to explain fome of the caufes which gave fuch energy to the ancient Mufic, and which ftill endear the melody of every country to its own inhabitants: Perhaps, for the reafons mentioned above, if we were to recover the Mufic which once had fo much power in the early periods of the Greek ftates, it might have no fuch charms for modern ears, as fome great admirers of antiquity imagine. Inftrumental Mufic indeed, unaccompanied with dance and fong, was never held in efteem till the later periods of antiquity; in which a general feparation of thefe arts took place. Plato * calls inftrumental Mufic an unmeaning thing, and an abufe of melody.

There is another caufe, which might probably contribute to make the

* De legibus.

the ancient Mufic more powerfully expreffive. In the infant ftate of focieties, * Men's feelings and paffions are ftrong, becaufe they are never difguifed nor reftrained; their imaginations are warm and luxuriant, from never having fuffered any check. This difpofes them to that enthufiafm fo favourable to Poetry and Mufic. The effufions of Genius among fuch a people may often poffefs the moft pathetic fublimity and fimplicity of ftile, though greatly deficient in point of elegance and regularity. And it is to be obferved, that thefe laft qualities are more peculiarly requifite in fome of the other fine arts, than they are in that fpecies of Mufic which is defigned to affect the paffions, where too much ornament is always hurtful; and in place of promoting, is much more likely to defeat the defired effect †.

The

---

* This fubject is treated with great accuracy and judgment by Dr. Blair, in his elegant differtation on the poems of Offian.

† Simplicity and concifenefs are never-failing characteriftics of the ftile of a fublime writer. He refts on the majefty of his fentiments, not on

the

The tranquillity too of rural life, and the variety of images with which it fills the imagination, have as beneficial an influence upon Genius, as they have upon the difpofitions of the heart. The country, and particularly the paftoral countries, are the favourite receffes of Poetry and Mufic.

The introduction of harmony opened a new world in Mufic. It promifed to give that variety which melody alone could never afford, and likewife to give melody an additional charm and energy. Unfortunately the firft compofers were fo immerfed in the ftudy of harmony,

which the pomp of his expreffions. The main fecret of being fublime, is to fay great things in few and plain words: for every fuperfluous decoration degrades a fublime idea. The mind rifes and fwells, when a lofty defcription or fentiment is prefented to it in its native form. But no fooner does the poet attempt to fpread out this fentiment or defcription, and to drefs it round and round with glittering ornaments, than the mind begins to fall from its high elevation; the tranfport is over; the beautiful may remain, but the fublime is gone. Dr. Blair's Critical Differtation on the poems of Offian.

The application of thefe ingenious obfervations to Mufic is too obvious to need any illuftration.

which soon appeared to be a science of great extent and intricacy, that these principal ends of it were forgot. They valued themselves on the laboured construction of parts which were multiplied in a surprising manner.—In fact, this art of counterpoint and complicated harmony, invented by Guido in the eleventh century, was brought to its highest degree of perfection by Palæstrini, who lived in the time of Leo X. But this species of Music could only be understood by the few who had made it their particular study. To every one else it appeared a confused jargon of sounds without design or meaning. To the very few who understood it there appeared an evident deficiency in air or melody, especially when the parts were made to run in strict fugues or canons, with which air is in a great measure incompatible.—Besides the real deficiency of air in these compositions, it required the attention to be constantly exerted to trace the subject of the Music, as it was alternately carried on through the several parts;

an attention inconsistent with what delights the ear, much more with what touches the passions; where this is the design of the Composer, the mind must be totally disengaged, must see no contrivance, admire no execution; but be open and passive to the intended impression.

We must however acknowledge, that there was often a Gravity, a Majesty, and Solemnity, in these old full compositions, admirably suited for the public services of the Church. Although perhaps less fitted to excite particular passions, yet they tended to sooth the mind into a tranquillity that disengaged it from all earthly cares and pleasures, and at the same time disposed it to that peculiar elevation which raises the soul to Heaven, especially when accompanied by the sweet and solemn notes of the Organ.

The artifice of fugues in vocal Music seems in a peculiar manner ill adapted to affect the passions. If every one of four voices is expressing a different sentiment and a different musical passage at the same time, the

hearer

hearer cannot poffibly attend to, and be affected by them all.—This is a ſtile of compoſition in which a perſon, without the leaſt Taſte or Genius, may become a confiderable proficient, by the mere force of ſtudy: But without a very great ſhare of theſe, to give ſpirit and meaning to the leading airs or ſubjects, ſuch compoſitions will always be dry and unaffecting. Catches, indeed, are a ſpecies of fugues, highly productive of mirth and jollity; but the pleaſure we receive from theſe ſeldom ariſes either from the melody itſelf, or from its being peculiarly expreſſive of the ſubject. It ariſes principally from the droll and unexpected aſſemblage of words from the different parts, and from the ſpirit and humour with which they are ſung.

Beſides the objections that lie againſt all complex Muſic with reſpect to its compoſition, there are others ariſing from the great difficulty of its execution. It is not eaſy to preſerve a number of inſtruments, playing together, in tune. Stringed inſtruments are falling, while wind inſtruments

ſtruments naturally riſe in their tone during the performance. It is not even ſufficient that all the performers play in the moſt exact tune and time. They muſt all underſtand the ſtile and deſign of the compoſition, and be able to make the reſponſes in the fugue with proper ſpirit. Every one muſt know how to carry on the ſubject with the proper expreſſion, when it is his turn to lead; and when he falls into an auxiliary part, he muſt know how to conduct his accompanyment in ſuch a manner as to give an additional force to the leading ſubject. But muſical taſte and judgment are moſt remarkably diſplayed in the proper accompanying of vocal Muſic, eſpecially with the thorough baſs. If this is not conducted with the ſtricteſt attention to heighten the intended expreſſion of the ſong, it deſtroys it altogether, as frequently happens from the throwing in the full chords, when a ſingle note ſhould only have been ſtruck, or when perhaps the accompanyment ſhould have ceaſed altogether.

These are difficulties few performers have an idea of, and fewer are able to conquer. Most of them think they sufficiently acquit themselves, if they play in tune and in time; and vanity often leads them to make their voice or instrument to be heard above the rest, without paying the least regard to the design of the Composer.

It has been much the fashion, for some years past, to regard air alone in musical compositions; and the full and regular works of harmony have fallen into neglect, being considered as cold and spiritless. This change has been introduced by composers, who unfortunately happened to be great performers themselves. These people had no opportunities, in the old compositions, of shewing the dexterity of their execution; the wild and extravagant flights which they indulged, in order to display this, being absolutely destructive of the harmony. They introduced therefore Solos of their own composition, or Concertos, which from the thinness and meagreness of the parts,

cannot

cannot be confidered in any other light than Solos. — It is not eafy to characterife the ftile of moft of thefe pieces. In truth they have no character or meaning at all. The authors of them are little concerned what fubject they choofe, their fingle view being to excite the furprife and admiration of their hearers. This they do by the moft unnatural and wild excurfions, that have not the remoteft tendency to charm the ear or touch the heart. In many paffages they are grating to the ear, when performed by the beft hands, but when executed by ordinary performers, they are perfectly intolerable. Thefe compofitions therefore want the merit which full harmony poffeffes, and are deficient in that fimplicity, fpirit, and elegance, which alone can recommend melody.

 The prefent mode is to admire a new noify ftile of compofition, lately cultivated in Germany, and to defpife Corelli as wanting fpirit and variety. The truth is, Corelli's ftile and this will not bear a comparifon. Corelli's excellence confifts in the chaftity chaſtity of his compoſition and in the richneſs and ſweetneſs of his harmonies. The other ſometimes pleaſes by its ſpirit and a wild luxuriancy, which makes an agreeable variety in a concert, but poſſeſſes too little of the elegance and pathetic expreſſion of Muſic to remain long the public Taſte. The great merit of that nobleman's compoſitions, who firſt introduced this ſpecies of Muſic into this country, and his own ſpirited performance of them, firſt ſeduced the public ear. They are certainly much ſuperior to any of the kind we have yet heard; though, by the delicacy of the airs in his ſlow movements, he diſplays a Genius capable of ſhining in a much ſuperior ſtile of Muſic.

Though Muſic, conſidered in its uſeful application, to delight the ear and touch the paſſions of the bulk of Mankind, requires the utmoſt ſimplicity, yet, conſidered as an art, capable of giving a laſting and varied pleaſure to the few, who from a ſtronger natural Taſte devote part of their time and attention to its cultivation,

tivation, it both admits, and requires variety, and even fome degree of complication.—Not only the ear but the mufical Tafte becomes more delicate by cultivation.

When the ear becomes acquainted with a variety of melodies, it begins by degrees to relifh others, befides thofe which are national. A national melody may have expreffions for only a few affections. A cultivated and enlarged Tafte eafily adopts a greater variety of expreffions for thefe and other affections, and learns, from the deepeft receffes of harmony, to exprefs fome that have never been excited by any national Mufic.

When one practifes Mufic much, the fimplicity of melody tires the ear. When he begins to hear an air he was formerly acquainted with, he immediately recollects the whole, and this anticipation often prevents his enjoying it. He requires therefore the affiftance of harmony, which, without hurting the melody, gives a variety to the Mufic, and fometimes renders the melody more
H 3 expreffive.

expreffive.— Practice enables one to trace the fubject of a complex Concerto, as it is carried through the feveral parts, which to a common ear is an unmeaning jumble of founds. Diftinct from the pleafure which the ear receives here from the Mufic, there is another, which arifes from the perception of the contrivance and ingenuity of the compofer.— This enjoyment, it muft be owned, is not of that heart-felt fort which fimple Mufic alone can give, but of a more fober and fedate kind, which proves of longer duration: And it muft be confidered, that whatever touches the heart or the paffions very fenfibly, muft be applied with a judicious and very fparing hand. — The fweeteft and fulleft chords muft be feldom repeated, otherwife the certain effects is fatiety and difguft.— They who are beft acquainted with the human heart, need not be told that this obfervation is not confined to Mufic.

On the whole we may obferve, that mufical Genius confifts in the invention of melody fuited to produce

duce a defired effect on the mind.—Mufical Tafte confifts in conducting the melody with fpirit and elegance, in fuch a manner as to produce this fingle effect in its full force.

Judgment in Mufic is fhewn in the contrivance of fuch harmonious accompanyments to the melody as may give it an additional energy, and a variety, without deftroying its fimplicity; in the preparation and refolution of difcords; and in the artful tranfitions from one key to another.—Tafte in a performer confifts in a knowledge of the compofer's defign, and expreffing it in a fpirited and pathetic manner, without any view of fhewing the dexterity of his own execution.

But though all thefe circumftances of compofition and performance fhould concur in any piece of Mufic, yet it muft always fail in affecting the paffions, unlefs its meaning and direction be afcertained by adapting it to fentiment and pathetic compofition.

It exerts its greateft powers when ufed as an affiftant to Poetry: hence the

the great superiority of vocal to instrumental Music, the human voice being capable of more justness, and at the same time of a more delicate musical expression, than any instrument whatever; the perfection of an instrument depending on its nearest approach to it. Vocal music is much confined by the language it is performed in. The harmony and sweetness of the Greek and Italian languages give them great advantages over the English and French, which are harsh, unmusical, and full of consonants; and this, among other inconveniences, occasions perpetual sacrifices of the quantity to the modulation.* This is one great cause of the slightness and want of variety of the French Music, which they in vain endeavour to cover and supply by laboured and complex accompanyments.

As vocal music is the first and most natural Music of every country, it is reasonable to expect it to bear some analogy to the Poetry of the country, to which it is always adapted.

* Rousseau.

adapted.—The remarkable superiority of the Scotch songs to the English, may in a great measure be accounted for from this principle. The Scotch songs are simple and tender, full of strokes of Nature and Passion. So is their Music. Many of the English songs abound in quaint and childish conceits. They all aim at wit, and sometimes attain it; but Music has no expression for wit, and the Music of their songs is therefore flat and insipid, and so little esteemed by the English themselves, that it is in a perpetual fluctuation, and has never had any characteristic stile. *

On the other hand, England has produced many admirable composers of Church Music. Their great attachment

---

* Dr. Brown very ingeniously observes, that most countries peopled by colonies, which, after a certain period of civilization, have issued from their native soil, possess no characteristic Music of their own; that the Irish, Welsh, and Scotch are strictly natives, and accordingly have a Music of their own: that the English, on the contrary, are a foreign mixture of late-established colonies, and, as a consequence of this, have no native Music; and that the original Music of England must be sought for in Wales.

tachment to Counterpoint hath indeed often led them into a wrong track; in other refpects, they have fhewn both Genius and Tafte.—Religion opens the ampleft field for mufical, as well as poetical Genius; it affords almoft all the variety of fubjects, which Mufic can exprefs; the fublime, the joyous, the chearful, the ferene, the devout, the plaintive, the forrowful. It likewife warms the heart with that enthufiafm fo peculiarly neceffary in all works of Genius. Accordingly our fineft compofitions in Mufic, are in the Church ftile. Handel, far advanced in life, when his conftitution and fpirits feemed nearly exhaufted, was fo roufed by this fubject, that he exhibited proofs of extent and fublimity of Genius in his Meffiah, fuperior to any he had fhewn in his moft vigorous period of life. We have another inftance of the fame kind in Marcello, a noble Venetian, who fet the firft fifty Pfalms to Mufic. In this work he has united the fimplicity and pathos of the ancient Mufic with the grace and variety of the modern.

modern. In compliance with the Taſte of the times he was ſometimes forced to leave that ſimplicity of ſtile which he loved and admired, but by doing ſo he has enriched the art with a variety of the moſt expreſſive and unuſual harmonies.

The great object in vocal Muſic is to make the Muſic expreſſive of the ſentiment. How little this is uſually regarded appears by the practice of ſinging all the parts of a ſong to the ſame Muſic, though the ſentiments and paſſions to be expreſſed be ever ſo different. If the Muſic has any character at all, this is a manifeſt violation of Taſte and common ſenſe, as it is obvious every different ſentiment and paſſion ſhould be expreſſed in a ſtile peculiarly ſuited to itſelf.

But the moſt common blunder, in compoſers, who aim at expreſſion, is their miſtaking imitation for it. —

* Muſic, conſidered as an imitative art, can imitate only ſounds or motion, and this laſt but very imperfectly. A compoſer ſhould make his

---

* See Harris and Aviſon.

his Music expressive of the sentiment, and never have a reference to any particular word used in conveying that sentiment, which is a common practice, and really a miserable species of punning. Besides, where imitation is intended, it should generally be laid upon the instrumental accompanyments, which by their greater compass and variety are fitter to perform the imitation, while the voice is left at liberty to express the sentiment. When the imitation is laid upon the voice, it obliges it to a strained and unnatural exertion, and prevents the distinct articulation of the words, which it is necessary to preserve, in order to convey the meaning of the song.— Handel sometimes observed this very carefully, at other times, as his Genius or attention was very unequal, he entirely neglected it. In that beautiful song of the Il Penserofo,

> ' Oft on a plat of rising ground,
> ' I hear the far-off curfew sound,'

he has thrown the imitation of the bell, with great art and success, into

to

to the symphony, and reserves the song entire for the expression of that pleasing tranquil melancholy, which the words so emphatically convey. He has shewn the same address in the celebrated song of Acis and Galatea,

> 'Hush, ye little warbling quire,'

where he has laid the imitation of the warbling of the birds upon the symphony and accompanyments, and preserves in the song that simplicity and languishing tenderness, which the subject of it particularly required. —On the other hand, in the song in Semele,

> 'The morning lark to mine accords his note,
> 'And tunes to my distress his warbling throat,'

he runs a long and laboured division on the word Warbling; and after all, the voice gives but a very faint imitation of the warlbing of the lark, though the violins in the symphony could have expressed it with great justness and delicacy.

In the union of Poetry and Music, the Music should be subservient to the

the Poetry: the very reverse is the common practice; the Poetry is ever made subordinate to the Music. Handel made those who composed the words of his Oratorios, alter and transpose them, as he thought best suited his Music: and as no Man of Genius could submit to this, we generally find the Poetry the most wretched imaginable.

We have frequently a more shocking instance of the little regard the composer has to the Poetry, and to the effect which should be left upon the Mind, in the unmeaning repetition of the first part of the Music after the second. It frequently happens, that a succession of very opposite passions takes place in the course of a song; for instance, from anger to reconciliation and tenderness, with which the sense requires it should conclude; yet the composer sometimes constructs his Music in such a way, as requires a return from the second to the first part with which the song must end. This is not only a glaring absurdity in point of sense, but distracts the Mind by
a most

a moſt unnatural ſucceſſion of paſſions.—

We have another inſtance of the little regard paid to the ultimate end of Muſic, the affecting the heart and paſſions, in the univerſally allowed practice of making a long flouriſh or cadence at the cloſe of a ſong, and ſometimes at other periods of it. In this the performer is left at liberty to ſhew the utmoſt compaſs of his throat and execution; and all that is required, is, that he ſhould conclude in the proper key; the performer accordingly takes this opportunity of ſhewing the audience the extent of his abilities, by the moſt fantaſtical and unmeaning extravagance of execution. The diſguſt which this gives to ſome, and the ſurprize which it excites in all the audience, breaks the tide of paſſion in the ſoul, and deſtroys all the effects which the compoſer has been ſtraining to produce.

It may be obſerved that the loud applauſe ſo frequently given to pieces of Muſic, ſeldom implies any compliment either to the compoſition itſelf

self or to the performer's juſt execution of it. They only expreſs our admiration of the performer's fine ſhake, or ſwelling of a note, his power of protracting a note twice as long as another could do without loſing his breath entirely, or of the variety of his cadence running out into the moſt extraneous modulation, and then artfully conducted to a proper concluſion in the key. But all theſe feats of art, the better they are executed, and the greater ſurprize they excite, the more effectually do they deſtroy the impreſſion of the preceding Muſic, if it was ever capable of producing any. They are in general as little eſſential to good Muſic, as the tricks of a Harlequin are to that gracefulneſs, elegance, and dignity of movement, which conſtitute the perfection of dancing. The genuine applauſe beſtowed on Muſic is to be ſought for in the profound ſilence, in the emphatic looks, and in the tears of the audience.

Our Oratorios labour under two diſadvantages; their being deprived of action and ſcenery; and their having

having no unity or defign as a whole. They are little elfe than a collection of fongs pretty much independent of one another. Now the effect of a dramatic performance does not depend on the effect of particular paffages, confidered by themfelves, but on that artful conftruction, by which one part gives ftrength to another, and gradually works the Mind up to thofe fentiments and paffions, which it was the defign of the author to produce.

The effects of Mufic depend upon many other circumftances befides its connection with Poetry. The effect, for inftance, of Cathedral Mufic depends greatly on its being properly adapted to the particular fervice of the day, and difcourfe of the preacher; and fuch a direction of it requires great tafte and judgment. Yet this is never attended to: the whole conduct of it is left to the caprice of the organift, who makes it airy or grave, chearful or plaintive, as it fuits his own fancy, and often degrades the folemnity and gravity
fuitable

suitable to divine worship, by the lightest and most trivial airs.

We see the same want of public Taste in the Music performed between the acts in * Tragedy, where the tone of passion is often broke in upon, and destroyed by airy and impertinent Music.

The effect of Music may sometimes be lost by an unhappy association of ideas with the person and character of a performer. When we hear at the Oratorio an Italian eunuch squeaking forth the vengeance of divine wrath, or a gay lively strumpet pouring forth the complaint of a deeply penitent and contrite heart, we must be hurt by such an association.

These observations relate principally to the public Taste of Music in Britain, if the public here can be said to have any Taste in this subject.

I shall readily allow that Music, considered merely as the art of affecting the ear agreeably by the power of sounds, is at present in a higher state than perhaps it has ever been

* Elements of Criticism.

been in any period; that the principles of harmony were never so well ascertained; and that there never was at any time so great a number of performers, in every branch of the art, distinguished for the spirit, brilliancy, and elegance of their execution. But notwithstanding all these advantages, it appears to be a fact, of which all men of common sense and observation, whether learned in the science or not, are equally judges; that Music, considered as the art of deeply affecting the heart, and commanding the passions by the power of sounds, is in a very low state, and that the principles on which these great and important effects depend, are either unknown or neglected. Of late years several composers of the highest rank seem to have been very sensible of this capital defect of our modern Music. In Italy particularly, that native country of all the elegant arts, a chastity, a simplicity and pathos of style has been cultivated by some eminent Masters, and successfully imitated by others in different parts of Europe.

rope. But the evil I complain of ſeems too complicated and too deeply rooted to admit now of a cure. The rage for variety is ſo exceſſive, and the Taſte, of courſe, ſo indiſcriminating, that compoſers and performers, who depend on the public for their ſubſiſtence, muſt ſatisfy it with any food they can procure, if it has only novelty to recommend it.

The wild effuſions of unbridled fancy, are often honoured with the titles of invention, ſpirit, and genius; and Taſte ſeems in general to mean nothing but an attachment to what is new, and a contempt for whatever is old in Muſic. Hence it ſeems to be now very generally admitted, that there are no fixt principles of Taſte in Muſic, as in the other fine Arts, and that it has no foundation but in caprice and faſhion. But I conceive that the principles of juſt Taſte in this Art, are as permanently founded in truth and human Nature, as thoſe of any art or ſcience whatever, and that the principles may be as certainly aſcertained by collecting and arranging the genuine feelings of
Nature.

Nature. The principles which deserve the chief attention, as being the first in point of dignity and utility, are those which relate to the power of Music, in commanding the passions; next to these, the principle of the art exercised merely with the view of amusement, by a transient gratification of the ear, should be examined and ascertained; and in the last and lowest place, the simple powers of execution may be considered as employed with the sole view of exciting surprize and admiration of the performer's abilities.

I could not pursue this subject farther without entering deeply into the intricacies of the technical part of Music, which I have carefully endeavoured to avoid. My design was only to shew, that Taste in Music has its foundation in Nature and common sense; that its noblest powers have been neglected, and that Men of sense and genius should not imagine they want an ear or a musical Taste, because they do not relish much of the modern Music, as in many

many cafes this is rather a proof of the goodnefs of both.

After all, it cannot be expected, that either Mufic or any of the fine arts, will ever be cultivated in fuch a manner as to make them ufeful and fubfervient to life, till the natural union be reftored which fo happily fubfifted between them and philofophy in ancient days; when philofophy not only gave to the world the moft accomplifhed generals and ftatefmen, but prefided with the greateft luftre and dignity over Rhetoric, Poetry, Mufic, and all the elegant arts that polifh and adorn Mankind.

SECTION IV.

It was formerly observed, that the pleasures arising from works of Taste and Imagination were confined to a small part of Mankind, and that although the foundations of a good Taste are laid in Human Nature, yet without culture it never becomes a considerable source of pleasure. As we formerly made a few observations on the the real effects produced by a cultivated Taste in some of the fine arts, we shall proceed to consider its influence on the pleasure arising from such works of Genius as are in a particular manner addressed to the Imagination and the Heart. This pleasure, in the earlier part of life, is often extremely high. Youth, indeed has peculiar advantages in this respect. The Imagination is then lively and vigorous,

gorous, the heart warm and feeling, equally open to the joyous impreffions of wit and humour, the force of the fublime, and every fofter and more delicate fentiment of humanity. It is matter of real concern to obferve the gradual decay of this innocent and rich fource of enjoyment, together with many others equally pure and natural.—Nature, it is true, has allotted different pleafures to different periods of life; but there is no reafon to think, that Nature has totally excluded any period from thofe pleafures of which we are now treating.

We have already lamented that many of the ufeful fciences as well as fine arts were left entirely in the hands of men unaffifted with learning and philofophy; but there is fome reafon to fufpect, that thefe affiftances have commonly been applied to works of Tafte and Imagination in fuch a manner, as has rather weakened than added to their force and influence.—This fubject is interefting, and deferves a particular difcuffion.

The

The Imagination, like every thing in nature, is subjected to general and fixt laws, which can only be discovered by experience. But it is no easy matter precisely to ascertain these laws. The subject is so fleeting, so various in different countries, in different constitutions of Men, and even in the same person in different periods and situations in life, that it requires the talents of a person of the most enlarged knowledge of Mankind, to reduce its laws to any kind of system; and this person likewise must be possessed of the most delicate sensibility of Heart and Imagination, otherwise he cannot understand what he is employed about.—Such a system of laws, particularly relating to dramatic and epic Poetry, was formed by some great men of antiquity, and has been since very universally adopted. Light has thereby been thrown on some of the great principles of criticism; and rules have been established, founded on the experience of such beauties as were discovered to please most universally. But without detracting from the merit of the ancient critics, it muft be obferved, that nothing tends more to check the improvement of any art or fcience, than the reducing all its principles too haftily into a regular fyftem. The bulk of Mankind are incapable of thinking or judging for themfelves on any fubject. There are a few leading fpirits whom the reft muft follow. This makes fyftems fo univerfally acceptable. If they cannot teach people to think and to feel, they teach them what to fay, which anfwers all the purpofes of the moft univerfally ruling paffion among Mankind, Vanity.

Thefe obfervations are particularly applicable to fyftems and rules of criticifm. When thefe are confidered as affiftances merely to the operations of Tafte; as giving proper openings for the difcernment of beauty, by collecting and arranging the feelings of Nature, they promote the improvement of the fine arts. But when they are confidered as fixed and eftablifhed ftandards, from which there lies no further appeal;
when

when they would impofe upon us the weight of authority, and fix a precife and narrow line, beyond which works of Imagination muft not ftray; in this cafe they do infinitely more harm than good. Tafte, of all the powers of the Mind, is leaft fuited to and moft impatient of fuch ftrict confinement. Some general principles may be pointed out, but to dream of applying always the fquare and the compafs to fuch thin and delicate feelings, as thofe of the Imagination, is a vain attempt. Add to this, that all criticifm muft, in a certain degree, be temporary and local.

Some tempers, and even fome nations, are moft pleafed with Nature in her faireft and moft regular forms, while others admire her in the great, the wonderful, and the wild. Thus elegance, regularity and fentiment are chiefly attended to in France, and French criticifm principally regards thefe; but its rules can with no propriety be applied in England, where the natural Genius or Tafte of the people is very different. The

grand,

grand, the fublime, the furprifing, and whatever very forcibly ftrikes the Imagination, ought there to be principally regarded. Where thefe are wanting, the utmoft elegance and propriety will appear cold and infipid: where thefe are found, elegance and propriety can be in a good meafure difpenfed with.

Whenever what is called a very correct Tafte generally prevails, the powers of Genius and Invention gradually languifh; and the conftant attention to prevent giving offence to a few, renders it impoffible to give much pleafure to any.

Refinement and delicacy of Tafte is an acquifition very dangerous and deceitful. It flatters our pride by giving us a confcious fuperiority over the reft of Mankind, and, by fpecious promifes of enjoyment unknown to vulgar Minds, often cheats us out of thofe pleafures which are equally attainable by the whole fpecies, and which Nature intended every one fhould enjoy. People poffeffed of extreme delicacy are haunted as it were with an evil Genius,

nius, by certain ideas of the coarse, the low, the vulgar, the irregular, which strike them in all the natural pleasures of life, and render them incapable of enjoying them.

There is scarcely an external or internal sense but may be brought by constant indulgence and attention, to such a degree of acuteness as to be disgusted at every object that is presented to it.—This extreme sensibility and refinement, though at first usually produced by vanity and affectation, yet by a constant attention to all the little circumstances that feed them, soon become real and genuine. But nature has set bounds to all our pleasures. We may enjoy them safely within these bounds, but if we refine too much upon them, the certain consequence is disappointment and chagrin.

When such a false delicacy, or, what has much the same effect, when the affectation of it becomes generally prevalent, it checks, in works of Taste, all vigorous efforts of Genius and Imagination, enervates the force of language, and produces that mediocrity,

mediocrity, that coldnefs and infipidity of compofition, which does not indeed greatly difguft, but never can give high pleafure. This is one bad effect of criticifm falling into wrong hands; efpecially when Men poffeffed of mere learning and abftract philofophy condefcend to beftow their attention on works of Tafte and Imagination. As fuch Men are fometimes deficient in thofe powers of Fancy, and that fenfibility of Heart, which are effential to the relifhing fuch fubjects, they are too often apt to defpife and condemn thofe things of which they have no right to judge, as they are neither able to perceive, nor to feel them.

A clear and acute Underftanding is far from being the only quality neceffary to form a perfect critic. The Heart is often more concerned here than the Head. In general, it feems the more proper bufinefs of true philofophical criticifm to obferve and watch the excurfions of fancy at a diftance, than to be continually checking all its little irregularities. Too much reftraint and
pruning

pruning is of more fatal confequence here than a little wildnefs and luxuriancy.

The * beauties of every work of Tafte are of different degrees, and fo are its blemifhes. The greateft blemifh is the want of fuch beauties as are charaƈteriftic, and effential to its kind. Thus in dramatic Poetry one part may be conftruƈted according to the laws of unity and truth, whilft another direƈtly contradiƈts them. The French, by their great attention to the general œconomy and unity of their fable, and the conftruƈtion of their fcenes, have univerfally obtained the charaƈter of fuperior correƈtnefs to the Englifh. Their reputation in this refpeƈt is well founded. In their dramatic writings we meet with much lefs that offends: and it muft alfo be acknowledged, that befides mere regularity of conftruƈtion, they poffefs in a high degree the merit of beautiful Poetry and tender fentiments. But when we examine them in another light, we find them excelled by the Englifh.

* Mufæum, Vol. I.

English. There is a want of force, often a degree of languor, even in their beſt pieces. The ſpeeches are generally too long and declamatory, the ſentiments too fine-ſpun, and the character enervated by a certain French appearance with which they are apt to be marked. Whereas, in the Engliſh theatre, if there be leſs elegance and regularity, there is more fire, more force, and more ſtrength. The paſſions ſpeak more their own native language; and the characters are drawn with a coarſer indeed, but however with a bolder hand.—Shakeſpeare, by his lively creative Imagination, his ſtrokes of Nature and Paſſion, and by preſerving the conſiſtency of his characters, amply compenſates for his tranſgreſſions againſt the rules of time and place, with which the Imagination can eaſily diſpenſe. His frequently breaking the tide of the Paſſions, by the introduction of low and abſurd comedy, is a more capital tranſgreſſion againſt Nature and the fundamental laws of the drama.

Probability is one of the boundaries,

ries, within which it has pleafed criticifm to confine the Imagination. This appears plaufible, but upon enquiry will perhaps be found too fevere a reftraint. It is obferved by the ingenious and elegant Author of the Adventurer, that events may appear to our reafon not only improbable, but abfurd and impoffible, whilft yet the Imagination may adopt them with facility and delight. The time was, when an univerfal belief prevailed of invifible agents interefting themfelves in the affairs of this world. Many events were fuppofed to happen out of the ordinary courfe of things by the fupernatural agency of thefe fpirits, who were believed to be of different ranks, and of different difpofitions towards Mankind. Such a belief was well adapted to to make a deep impreffion on fome of the moft powerful principles of our Nature, to gratify the natural paffion for the marvellous, to dilate the Imagination, and to give boundlefs fcope to its excurfions.

 In thofe days the old Romance was in its higheft glory. And though

I 5   a belief

a belief of the interpofition of thefe invifible powers in the ordinary affairs of Mankind has now ceafed, yet it ftill keeps its hold of the Imagination, which has a natural propenfity to embrace this opinion. Hence we find that Oriental tales continue to be univerfally read and admired, by thofe who have not the leaft belief in the Genii, who are the moft important agents in the ftory. All that we require in thefe works of Imagination is an unity and confiftency of character.* The Imagination willingly allows itfelf to be deceived into a belief of the exiftence of beings, which reafon fees to be ridiculous; but then every event muft take place in fuch a regular manner as may be naturally expected from the interpofition of fuch fuperior intelligence and power. It is not a fingle violation of truth and probability that offends, but fuch a violation as perpetually recurs. We have a ftrong evidence of the facility with which the Imagination is deceived, in the effects produced by a well-

* Adventurer.

well-acted Tragedy. The Imagination there soon becomes too much heated, and the Paſſions too much intereſted, to permit reaſon to reflect that we are agitated with the feigned diſtreſs of people entirely at their eaſe. We ſuffer ourſelves to be tranſported from place to place, and believe we are hearing the private ſoliloquy of a perſon in his chamber, while he is talking on a ſtage ſo as to be heard by thouſands.

The deception in our modern Novels is more perfect than in the old Romance; but as they profeſs to paint Nature and Characters as they really are, it is evident that the powers of fancy cannot have the ſame play, nor can the ſucceſſion of incidents be ſo quick nor ſo ſurprizing. It requires therefore a Genius of the firſt claſs to give them that ſpirit and variety ſo neceſſary to captivate the Imagination, and to preſerve them from ſinking into dry narrative and tireſome declamation.

Notwithſtanding the ridiculous extravagance of the old Romance in many particulars, it ſeems calculated
to

to produce more favourable effects on the morals of Mankind, than our modern Novels.—If the former did not reprefent Men as they really are, it reprefented them as they ought to be; its heroes were patterns of courage, generofity, truth, humanity, and the moſt exalted virtues. Its heroines were diſtinguiſhed for modeſty, delicacy, and the utmoſt dignity of manners.—The latter reprefent Mankind too much what they are, paint fuch fcenes of pleafure and vice as ought never to fee the light, and thus in a manner hackney youth in the ways of wickednefs, before they are well entered into the world; expofe the fair fex in the moſt wanton and fhamelefs manner to the eyes of the world, by ſtripping them of that modeſt referve, which is the foundation of grace and dignity, the veil with which Nature intended to protect them from too familiar an eye, in order to be at once the greateſt incitement to love and the greateſt fecurity to virtue.—
In fhort, the one may miflead the Imagination; the other tends to inflame

flame the Paffions and to corrupt the heart.

The pleafure which we receive from Hiftory arifes in a great meafure from the fame fource with that which we receive from Romance. It is not the bare recital of facts that gives us pleafure. They muft be facts that give fome agitation to the Mind by their being important, interefting, or furprizing. But events of this kind do not very frequently occur in Hiftory, nor does it defcend to paint thofe minute features of particular perfons which are more likely to engage our affections and intereft our paffions than the fate of nations. It is not therefore furprizing that we find it fo difficult to keep attention awake in reading Hiftory, and that fewer have fucceeded in this kind of compofition than in any other. To render Hiftory pleafing and interefting it is not enough that it be ftrictly impartial, that it be written with the utmoft elegance of language, and abound in the moft judicious and uncommon obfervations. We are never agreeably

ably interested in a History, till we contract an attachment to some public and important cause, or some distinguished characters which it represents to us. The fate of these engages the attention and keeps the Mind in an anxious yet pleasing suspence. Nor do we require the author to violate the truth of History, by representing our favourite cause or hero as perfect; we will allow him to represent all their weaknesses and imperfections, but still it must be with such a tender and delicate hand as not to destroy our attachment. There is a sort of unity or consistency of character that we expect even in History. An author of any ingenuity can, if he pleases, easily disappoint this expectation, without deviating from truth. There are certain features in the greatest and worthiest Men, which may be painted in such a light as to make their characters appear little and ridiculous. Thus if an Historian be constantly attentive to check admiration, it is certainly in his power: but if the Mind be thus continually disappointed,

pointed, and can never find an object that may be contemplated with pleasure, though we may admire his Genius, and be instructed by his History, he will never leave a pleasing and grateful impression on the Mind. Where this is the prevailing spirit and genius of a History, it not only deprives us of a great part of the pleasure we expected from it, but leaves disagreeable effects on the Mind, as it stifles that noble enthusiasm, which is the foundation of all great actions, and produces a fatal scepticism, coldness, and indifference about all characters and principles whatsoever. We acknowledge indeed that this manner of writing may be of great service in correcting the narrow prejudices of party and faction; as they will be more influenced by the representations of one who seems to take no side, than by any thing which can be said by their antagonists.

But the principal and most important end of History, is to promote the interests of Liberty and Virtue, and

and not merely to gratify curiofity. Impartial Hiftory will always be favourable to thefe interefts. The elegance of its ftile and compofition, is chiefly to be valued, as it ferves to engage the reader's attention. But if an Hiftorian has no regard to what we here fuppofe fhould be the ultimate ends of Hiftory, if he confiders it only as calculated to give an exercife and amufement to the Mind, he may undoubtedly make his work anfwer a very different purpofe. The circumftances that attend all great events are fo complicated, and the weakneffes and inconfiftencies of every human character, however exalted and amiable, are fo various, that an ingenious writer has an opportunity of placing them in a point of view that may fuit whatever caufe he chufes to efpoufe. Under the fpecious pretence of a regard to truth, and a fuperiority to vulgar prejudices, he may render the beft caufe doubtful, and the moft refpectable character ambiguous. This may be eafily done without any abfolute deviation from Truth; by only fup-
preffing

pressing some circumstances, and giving a high colouring to others; by taking advantage of the frivolous and dissolute spirit of the age, which delights in seeing the most sacred and important subjects turned into ridicule; and by insinuations that convey, in the strongest manner, sentiments which the Author, from affected fear of the laws, or a pretended delicate regard to established opinions, seems unwilling fully and clearly to express. Of all the methods that have been used to shake those principles on which the virtue, the liberties, and the happiness of Mankind depend, this is the most dangerous as well as the most illiberal and disingenuous. It is impossible to confute a hint, or to answer an objection that is not fully and explicitly stated. There is a certain species of impartiality with which no man, who has good principles, or a sensible heart, will sit down to write History; that impartiality, which supposes an absolute indifference to whatever may be its consequences or the minds of the readers. Such an indifference,

indifference, in regard to the refult of our enquiries, is natural and proper in the abftract Sciences, and in thofe Philofophical difquifitions, where truth is the fingle and ultimate object, not connected with any thing that may engage the affections or effentially affect the interefts of Mankind. But a candid Hiftorian, who is the friend of Mankind, will difclaim this coldnefs and infenfibility: He will openly avow his attachment to the caufe of liberty and virtue, and will confider the fubferviency of his Hiftory to their interefts as its higheft merit and honour. He will be perfuaded that Truth, that impartial Hiftory, can never hurt thefe facred interefts; but he will never pretend fo far to diveft himfelf of the feelings of a Man, as to be indifferent whether they do or not.

A lively Imagination, and particularly a poetical one, bears confinement no where fo ill as in the ufe of Metaphor and Imagery. This is the peculiar province of the Imagination. The foundeft head can neither

ther affift nor judge in it. The Poet's eye, as it *glances from heaven to earth, from earth to heaven*, is ftruck with numberlefs fimilitudes and analogies, that not only pafs unnoticed by the reft of Mankind, but cannot even be comprehended when fuggefted to them. There is a correfpondence between certain external forms of Nature, and certain affections of the Mind, that may be felt, but cannot always be explained. Sometimes the affociation may be accidental, but it often feems to be innate. Hence the great difficulty of afcertaining the true fublime. It cannot in truth be confined within any bounds; it is entirely relative, depending on the warmth and livelinefs of the Imagination, and therefore different in different countries. For the fame reafon, wherever there is great richnefs and profufion of Imagery, which in fome fpecies of Poetry is a principal beauty, there are always very general complaints of obfcurity, which is increafed by thofe fudden tranfitions that bewilder

* Shakefpeare.

wilder a common reader, but are eaſily traced by a poetical one. An accurate ſcrutiny into the propriety of Images and Metaphors is fruitleſs. If it be not felt at firſt, it can ſeldom be communicated: while we endeavour to analyſe it, the impreſſion vaniſhes. The ſame obſervation may be applied to Wit, which conſiſts in a quick and unexpected aſſemblage of ideas, that ſtrike the Mind in an agreeable manner either by their reſemblance or their incongruity. Neither is the juſtneſs of humour a ſubject that will bear reaſoning. This conſiſts in a lively painting of thoſe weakneſſes of character, which are not of importance enough to raiſe pity or indignation, but only excite mirth and laughter. One muſt have an idea of the original to judge of, or be affected by the repreſentation, and if he does not ſee its juſtneſs at the firſt glance, he never ſees it. For this reaſon moſt works of humour, ridicule, and ſatire, which paint the particular features and manners of the times, being local and tranſient, quickly

quickly lofe their poignancy, and become obfcure and infipid.

Whatever is the object of Imagination and Tafte can only be feen to advantage at a certain diftance, and in a particular light. If brought too near the eye, the beauty which charmed before appears faded, and often diftorted. It is therefore the bufinefs of judgment to afcertain this point of view, to exhibit the object to the Mind in that pofition which gives it moft pleafure, and to prevent the Mind from viewing it in any other. This is generally very much in our own power. It is an art which we all practife in common life. We learn by habit to turn to the eye the agreeable fide of any object which gives us pleafure, and to keep the dark one out of fight. If this be kept within any reafonable bounds, the foundeft judgment will not only connive at, but approve it. — Whatever we admire or love, as great, or beautiful, or amiable, has certain circumftances belonging to it, which, if attended to, would poifon our enjoyment. — We are agreeably
ftruck

struck with the grandeur and magnificence of Nature in her wildeſt forms, with the proſpect of vaſt and ſtupendous mountains; but is there any neceſſity for our attending, at the ſame time to the bleakneſs, the coldneſs, and the barrenneſs, which are univerſally connected with them? When a lover contemplates with rapture the charms of beauty and elegance, that captivate his heart, need he at the ſame time reflect how uncertain and tranſient the object of his paſſion is, and that the ſucceſſion of a few years muſt lay it mouldering in the duſt?

But we not only think it unneceſſary always to ſee the whole truth, but frequently allow and juſtify ourſelves in viewing things magnified beyond the truth. We indulge a manifeſt partiality to our friends, to our children, and to our native country. We not only keep their failings, as much as prudence will juſtify, out of ſight, but we exalt in our Imagination all their good qualities beyond their juſt value. Nor does the general ſenſe of Mankind condemn

condemn this indulgence; for this very good reason, because it is natural, and because we could not forego it, without losing at the same time all sense of friendship, natural affection, and patriotism. — There appears no sufficient reason why this conduct, which we observe in common life, should not be followed in our enquiries into works of Imagination. A person of a cultivated Taste, while he resigns himself to the first impressions of pleasure excited by real excellence, can at the same time, with the slightest glance of the eye, perceive whether the work will bear a nearer inspection. If it can bear this, he has an additional pleasure, arising from those latent beauties which strike the Imagination less forcibly. If he finds they cannot bear this examination, he should remove his attention immediately, and he should gratefully enjoy the pleasure he has already received.

A correct Taste is very much offended with Dr. Young's Night Thoughts; it observes that the representation there given of Human Life

Life is falfe and gloomy; that the poetry fometimes finks into childifh conceits or profaic flatnefs, but oftner rifes into the turgid or falfe fublime; that it is perplexed and obfcure; that the reafoning is often weak; and that the general plan of the work is ill laid, and not happily conducted.— Yet this work may be read with very different fentiments. It may be found to contain many touches of the moft fublime Poetry that any language has produced, and to be full of thofe pathetic ftrokes of Nature and Paffion, which touch the heart in the moft tender and affecting manner.—

Befides, the Mind is fometimes in a difpofition to be pleafed only with dark views of Human Life.

There are afflictions too deep to bear either reafoning or amufement. They may be foothed, but cannot be diverted. The gloom of the Night Thoughts perfectly correfponds with this ftate of Mind. It indulges and flatters the prefent paffion, and at the fame time prefents thofe motives of confolation

which

which alone can render certain griefs fupportable.—We may here obferve that fecret and wonderful endearment, which Nature has annexed to all our fympathetic feelings. We enter into the deepeft fcenes of diftrefs and forrow with a melting foftnefs of Heart, far more delightful than all the joys which diffipated and unthinking mirth can infpire. * Dr. Akenfide defcribes this very pathetically.

———————— Afk the faithful youth,
Why the cold urn of her, whom long he loved,
So often fills his arms; fo often draws
His lonely footfteps at the filent hour,
To pay the mournful tribute of his tears?
Oh! he will tell thee, that the wealth of worlds
Should ne'er feduce his bofom to forego
That facred hour, when ftealing from the noife
Of care and envy, fweet remembrance fooths
With virtue's kindeft looks his aking breaft,
And turns his tears to rapture.

He afterwards proceeds to paint, with all the enthufiafm of liberty and poetic Genius, and in all the fweetnefs and harmony of numbers, thofe heart-ennobling forrows, which the Mind feels by the reprefentation of the

* Pleafures of Imagination.

the prefent miferable condition of thofe countries, which were once the happy feats of Genius, Liberty, and the greateft virtues that adorn humanity.

What ought chiefly to be regarded in the culture of Tafte is to difcover thofe many beauties, in the works of Nature and Art, which would otherwife efcape our notice. Thomfon, in that beautiful defcriptive poem, the Seafons, pleafes from the juftnefs of his painting; but his greateft merit confifts in impreffing the Mind with numberlefs beauties of Nature, in her various and fucceffive forms, which formerly paffed unheeded. — This is the moft pleafing and ufeful effect of criticifm; to difplay new fources of pleafure unknown to the bulk of Mankind; and it is only fo far as it difcovers thefe, that Tafte can with reafon be accounted a bleffing.

It has been often obferved that a good Tafte and a good Heart commonly go together. But that fort of Tafte, which is conftantly prying into blemifhes and deformity, can have

have no good effect either on the Temper or the Heart. The Mind naturally takes a taint from thofe objects and purfuits in which it is ufually employed. Difguft, often recurring, fpoils the Temper, and a habit of nicely difcriminating, when carried into real life, contracts the Heart, and, by holding up to view the faults and weakneffes infeparable from every character, not only checks all the benevolent and generous affections, but ftifles all the pleafing emotions of love and admiration.

The habit of dwelling too much on what is ridiculous in fubjects of Tafte, when transferred into life, has likewife a bad effect upon the character, if not foftened by a large portion of humanity and good humour, as it confers only a fullen and gloomy pleafure, by feeding the worft and moft painful feelings of the human heart, envy and malignity. But an intimate acquaintance with the works of Nature and Genius, in their moft beautiful and amiable forms, humanizes and fweetens the Temper, opens and extends the Imagination, and difpofes to the moft pleafing views of Mankind and Providence. By confidering Nature in this favourable point of view, the Heart is dilated, and filled with the moft benevolent fentiments; and then indeed the fecret fympathy and connection between the feelings of Natural and Moral Beauty, the connection between a good Tafte and a good Heart, appears with the greateft luftre.

SECTION V.

WE proceed now to consider that principle of Human Nature which seems in a peculiar manner the characteristic of the species, the Sense of Religion. It is not my intention here to consider the evidence of religion as founded in truth; I propose only to examine it as a principle founded in Human Nature, and the influence it actually has, or may have, on the happiness of Mankind. —— The beneficial consequences which should naturally result from this principle, seem very obvious. There is something peculiarly soothing and comfortable in a firm belief that the whole frame of Nature is supported and conducted by an eternal and omnipotent Being, of infinite goodness, who intends, by the whole course of his providence, to promote the greateſt good of all his creatures; a belief that we are acquainted with the means of conciliating the Divine favor, and that in conſequence of this we have it in our own power to obtain it; a belief that this life is but the infancy of our exiſtence, that we ſhall ſurvive the ſeeming deſtruction of our preſent frame, and have it in our power to ſecure our entrance on a new ſtate of eternal felicity. If we believe that the conduct which the Deity requires of us is ſuch as moſt effectually ſecures our preſent happineſs, together with the peace and happineſs of ſociety, we ſhould of courſe conclude that theſe ſentiments would be fondly cheriſhed and adopted by all wiſe and good Men, whether they were ſuppoſed to ariſe from any natural anticipation of the Human Mind, the force of Reaſon, or an immediate revelation from the Supreme Being.

But though the belief of a Deity and of a future ſtate of exiſtence have univerſally prevailed in all ages and nations, yet it has been diverſified and connected with a variety of
ſuperſtitions,

superstitions, which have often rendered it useless, and sometimes hurtful to the general interests of Mankind. The Supreme Being has sometimes been represented in such a light, as made him rather an object of terror than of love; as executing both present and eternal vengeance on the greatest part of the world, for crimes they never committed, and for not believing doctrines which they never heard. — Men have been taught that they did God acceptable service by abstracting themselves from all the duties they owed to society, by denying themselves all the pleasures of life, and even by voluntarily enduring and inflicting on themselves the severest tortures which Nature could support. They have been taught that it was their duty to persecute their fellow-creatures in the most cruel manner, in order to bring them to an uniformity with themselves in religious opinions; a scheme equally barbarous and impracticable. In fine, Religion has often been used as an engine to deprive Mankind of their most valuaable privileges, and to subject them to the most despotic tyranny.

These pernicious consequences have given occasion to some ingenious Men to question, whether Atheism or superstition were most destructive to the happiness of society; while others have been so much impressed by them, that they seemed to entertain no doubt of its being safer to divest Mankind of all religious opinions and restraints whatever, than to run the risk of the abuses which they thought almost inseparable from them.—This seems to be the most favorable construction that can be put on the conduct of the patrons of Atheism. But however specious this pretence might have been some centuries ago, there does not at this time appear to be the least foundation for it. Experience has now shewn that Religion may subsist in a public establishment, divested of that absurd and pernicious Superstition which was only adventitious, and most apparently contrary to its genuine and original spirit and genius.—To separate Religion

gion entirely from Superstition, in every individual, may indeed be impossible, because it is impossible to make all Mankind think wisely and properly on any one subject, where the Understanding alone is concerned, much more where the Imagination and the affections are so deeply interested. But if the positive advantages of Religion to Mankind be evident, this should seem a sufficient reason for every worthy Man to support its cause, and at the same time to keep it disengaged from those accidental circumstances that have so highly dishonoured it.

Mankind certainly have a sense of right and wrong, independent of religious belief; but experience shews, that the allurements of present pleasure, and the impetuosity of passion are sufficient to prevent Men from acting agreeably to this moral sense, unless it be supported by Religion, the influence of which upon the Imagination and Passions, if properly directed, is extremely powerful.

We shall readily acknowledge that many of the greatest enemies of Religion have been distinguished for their honour, probity, and good nature. But it is to be considered, that many virtues as well as vices are constitutional. A cool and equal Temper, a dull Imagination, and unfeeling Heart, ensure the possession of many virtues, or rather are a security against many vices. They may produce temperance, chastity, honesty, prudence, and a harmless, inoffensive, behaviour. Whereas keen passions, a warm Imagination, and great sensibility of Heart, lay a natural foundation for prodigality, debauchery, and ambition; attended, however, with the seeds of all the social and most heroic virtues. Such a temperature of Mind carries along with it a check to its constitutional vices, by rendering those possessed of it peculiarly susceptible of religious impressions. They often appear indeed to be the greatest enemies to Religion, but that is entirely owing to their impatience of its restraints. Its most dangerous enemies

enemies have ever been among the temperate and chaste philosophers, void of passion and sensibility, who had no vicious appetites to be restrained by its influence, and who were equally unsusceptible of its terrors or its pleasures. Absolute Infidelity or settled Scepticism in Religion we acknowledge is no proof of want of Understanding, or a vicious disposition, but is certainly a very strong presumption of the want of Imagination and sensibility of Heart, and of a perverted Understanding. Some philosophers have been Infidels, few Men of taste and sentiment. Yet the examples of Lord Bacon, Mr. Locke, and Sir Isaac Newton, among many other first names in philosophy, are a sufficient evidence that religious belief is perfectly compatible with the clearest and most enlarged Understanding.

Several of those who have surmounted what they call religious prejudices themselves, affect to treat such as are not ashamed to avow their regard to Religion, as Men of
weak

weak Underftandings and feeble Minds. But this fhews either want of candor or great ignorance of Human Nature. The fundamental articles of Religion have been very generally believed by Men the moft diftinguifhed for acutenefs and accuracy of judgment. Nay, it is unjuft to infer the weaknefs of a perfon's head on other fubjects from his attachment even to the fooleries of Superftition. Experience fhews that when the Imagination is heated, and the affections deeply interefted, they level all diftinctions of Underftanding; yet this affords no prefumption of a fhallow judgment in fubjects where the Imagination and Paffions have no influence.

Feeblenefs of Mind is a reproach frequently thrown, not only upon fuch as have a fenfe of Religion, but upon all who poffefs warm, open, chearful Tempers, and Hearts peculiarly difpofed to love and friendfhip. But the reproach is ill founded. Strength of Mind does not confift in a peevifh Temper, in a hard inflexible Heart, and in bidding

ding defiance to God Almighty. It confifts in an active refolute Spirit, in a fpirit that enables a Man to act his part in the world with propriety, and to bear the misfortunes of life with uniform fortitude and dignity. This is a ftrength of Mind which neither Atheifm nor univerfal Scepticifm will ever be able to infpire. On the contrary, their tendency will be found to chill all the powers of Imagination; to deprefs Spirit as well as Genius; to four the Temper and contract the Heart. The higheft religious fpirit, and veneration for Providence breathes in the writings of the ancient Stoics; a fect diftinguifhed for producing the moft active, intrepid, virtuous Men that ever did honour to Human Nature.

Can it be pretended that Atheifm or Univerfal Scepticifm have any tendency to form fuch characters? Do they tend to infpire that magnanimity and elevation of Mind, that fuperiority to felfifh and fenfual gratifications, that contempt of danger and of death, when the caufe of virtue, of liberty, or their country require

quire it, which diftinguifh the characters of Patriots and Heroes? or is their influence more favorable on the humbler and gentler virtues of private and domeftic life? Do they foften the heart, and render it more delicately fenfible of the thoufand namelefs duties and endearments of a Hufband, a Father, or a Friend? Do they produce that habitual ferenity and chearfulnefs of temper, that gaiety of heart, which makes a Man beloved as a Companion? or do they dilate the heart with the liberal and generous fentiments, and that love of human kind, which would render him revered and bleffed as the patron of depreffed merit, the friend of the widow and orphan, the refuge and fupport of the poor and the unhappy?

The general opinion of Mankind, that there is a ftrong connection between a religious difpofition and a feeling Heart, appears from the univerfal diflike, which all Men have to Infidelity in the fair fex. We not only look on it as removing the principal fecurity we have for their virtue,

tue, but as the strongest proof of their want of that softness and delicate sensibility of Heart, which peculiarly endears them to us, and more effectually secures their empire over us, than any quality they can possess.

There are indeed some Men who can persuade themselves, that there is no Supreme Intelligence who directs the course of Nature; who can see those they have been connected with by the strongest bonds of Nature and Friendship gradually disappearing; who are persuaded that this separation is final and eternal, and who expect that they themselves shall soon sink down after them into nothing; and yet such Men appear easy and contented. But to a sensible Heart, and particularly to a Heart softened by past endearments of Love or Friendship, such opinions are attended with gloom inexpressible; they strike a damp into all the pleasures and enjoyments of life, and cut off those prospects which alone can comfort the soul under certain distresses,

distresses, where all other aid is feeble and ineffectual.

Scepticism, or suspence of judgment as to the truth of the great articles of Religion, is attended with the same fatal effects. Wherever the affections are deeply interested, a state of suspence is more intolerable, and more distracting to the Mind, than the sad assurance of the evil which is most dreaded.

There are many who have past the age of Youth and Beauty, and who have resigned the pleasures of that smiling season; who begin to decline into the vale of Years, impaired in their Health, depressed in their Fortunes, stript of their Friends, their Children, and perhaps, still more tender and endearing connections. What resource can this world afford them? It presents a dark and dreary waste, thro' which there does not issue a single ray of comfort. Every delusive prospect of Ambition is now at an end; long experience of Mankind, an experience very different from what the open and generous soul of youth had fondly dreamt of, has

has rendered the Heart almoſt inacceſſible to new Friendſhips. The principal ſources of Activity are taken away, when thoſe for whom we labour are cut off from us, thoſe who animated, and thoſe who ſweetened all the toils of life. Where then can the ſoul find refuge, but in the boſom of Religion? There ſhe is admitted to thoſe proſpects of Providence and Futurity, which alone can warm and fill the Heart. I ſpeak here of ſuch as retain the feelings of Humanity, whom Misfortunes have ſoftened and perhaps rendered more delicately ſenſible; not of ſuch as poſſeſs that ſtupid Inſenſibility which ſome are pleaſed to dignify with the name of Philoſophy.

It ſhould therefore be expected that thoſe Philoſophers, who ſtand in no need themſelves of the aſſiſtance of Religion to ſupport their virtue, and who never feel the want of its conſolations, would yet have the humanity to conſider the very different ſituation of the reſt of Mankind; and not endeavour to deprive them of what Habit, at leaſt, if they will

will not allow it to be Nature, has made neceſſary to their morals and to their happineſs. — It might be expected that Humanity would prevent them from breaking into the laſt retreat of the unfortunate, who can no longer be objects of their envy or reſentment, and tearing from them their only remaining comfort. The attempt to ridicule Religion may be agreeable to ſome, by relieving them from a reſtraint upon their pleaſures, and may render others very miſerable, by making them doubt thoſe truths, in which they were moſt deeply intereſted; but it can convey real good and happineſs to no one individual.

To ſupport openly and avowedly the cauſe of Infidelity may be owing in ſome to the vanity of appearing wiſer than the reſt of Mankind; to Vanity, that amphibious paſſion that ſeeks for food, not only in the affectation of every beauty, and every virtue that adorn Humanity, but of every vice and perverſion of the Underſtanding, that diſgrace it. The zeal of making proſelytes to it may often

often be attributed to a like vanity of poffeffing a direction and afcendency over the Minds of Men, which is a very flattering fpecies of fuperiority. But there feems to be fome other caufe that fecretly influences the conduct of fome that reject all Religion, who from the reft of their character cannot be fufpected of vanity, or any ambition of fuch fuperiority. This we fhall attempt to explain.

The very differing in opinion, upon any interefting fubject, from all around us, gives a difagreeable fenfation. This muft be greatly increafed in the prefent cafe, as the feeling, which attends Infidelity or Scepticifm in Religion, is certainly a comfortlefs one, where there is the leaft degree of fenfibility.—Sympathy is much more fought after by an unhappy Mind, than by one chearful and at eafe. We require a fupport in the one cafe, which in the other is not neceffary. A perfon therefore void of Religion feels himfelf as it were alone in the midft of fociety; and though for prudential reafons he choofes chooses on some occasions to disguise his sentiments, and join in some form of religious worship, yet this to a candid and ingenuous Mind must always be very painful; nor does it abate the disagreeable feeling which a social spirit has in finding itself alone and without any friend to sooth and participate its uneasiness. This seems to have a considerable share in that anxiety which Free-thinkers generally discover to make proselytes to their opinions, an anxiety much greater than what is shewn by those, whose Minds are at ease in the enjoyment of happier prospects.

The excuse, which these gentlemen plead for their conduct, is a regard for the cause of truth. But this is a very insufficient one. None of them act upon this principle, in its largest extent and application, in common life. Nor could any Man live in the world and pretend so to do. In the pursuit of happiness, * *our being's end and aim*, the discovery of truth is far from being the most important object. It is true the Mind

* Pope.

Mind receives a high pleasure from the investigation and discovery of Truth, in the abstract sciences, in the works of Nature and Art, but in all subjects, where the Imagination and Affections are deeply concerned, we regard it only so far as it is subservient to them.—One of the first principles of society, of decency, and of good manners, is, that no Man is entitled to say every thing he thinks true, when it would be injurious or offensive to his neighbour. If it was not for this principle, all Mankind would be in a state of hostility.

Suppose a person to lose an only child, the sole comfort and happiness of his life. When the first overflowings of Nature are past, he recollects the infinite goodness and impenetrable wisdom of the Disposer of all events, he is persuaded that the revolution of a few years will again unite him to his child never more to be separated. With these sentiments he acquiesces with a melancholy yet pleasing resignation to the Divine will. Now supposing all
this

this to be a deception, a pleafing dream, would not the general fenfe of Mankind condemn the Philofopher as barbarous and inhuman, who fhould attempt to wake him out of it?— Yet fo far does vanity prevail over good-nature, that we frequently fee Men, on other occafions of the moft benevolent Tempers, labouring to cut off that hope, which can alone chear the Heart under all the preffures and afflictions of Human Life, and enable us to refign it with chearfulnefs and dignity.

Religion may be confidered in three different views. Firft, As containing doctrines relating to the being and perfections of God, his moral adminiftration of the world, a future ftate of exiftence, and particular communications to Mankind by an immediate fupernatural revelation. —Secondly, As a rule of life and manners.—Thirdly, As the fource of certain peculiar affections of the Mind, which either give pleafure or pain, according to the particular genius and fpirit of the Religion that infpires them.

In

In the firſt of theſe views, which gives a foundation to all religious belief, and on which the other two depend, Reaſon is principally concerned. On this ſubject the greateſt efforts of human genius and application have been exerted, and with the moſt deſirable ſuccefs in thoſe great and important articles that ſeem moſt immediately to affect the intereſt and happineſs of Mankind. But when our enquiries here are puſhed to a certain length, we find that Providence has ſet bounds to our Reaſon, and even to our capacities of apprehenſion. This is particularly the caſe, with reſpect to infinity and the moral œconomy of the Deity. The objects are here in a great meaſure beyond the reach of our conception; and induction from experience, on which all our other reaſonings are founded, cannot be applied to a ſubject altogether diſſimilar to any thing we are acquainted with.—Many of the fundamental articles of Religion are ſuch, that the Mind may have the fulleſt conviction of their truth, but they muſt
be

be viewed at a diſtance, and are rather the objects of ſilent and religious veneration, than of metaphyſical diſquiſition. If the Mind attempts to bring them to a nearer view, it is confounded with their ſtrangeneſs and immenſity.

When we purſue our enquiries into any part of Nature, beyond certain bounds, we find ourſelves involved in perplexity and darkneſs. But there is this remarkable difference between theſe and religious enquiries: In the inveſtigation of Nature, we can always make a progreſs in knowledge, and approximate to the truth by the proper exertion of genius and obſervation; but our enquiries into religious ſubjects, are confined within very narrow bounds; nor can any force of reaſon or application lead the Mind one ſtep beyond that impenetrable gulf, which ſeparates the viſible, and inviſible world.

Though the articles of religious belief, which fall within the comprehenſion of Mankind, and ſeem eſſential to their happineſs, are few and

and simple, yet ingenious Men have contrived to erect them into most tremendous systems of metaphysical subtlety, which will long remain monuments both of the extent, and the weakness of human Understanding. The pernicious consequences of such systems, have been various. By attempting to establish too much, they have hurt the foundation of the most interesting principles of Religion.—Most Men are educated, in a belief of the peculiar, and distinguishing opinions of some one religious sect or other. They are taught that all these are equally founded on Divine authority, or the clearest deductions of Reason. By which means, their system of Religion hangs so much together, that one part cannot be shaken, without endangering the whole. But wherever any freedom of enquiry is allowed, the absurdity of some of these opinions, and the uncertain foundation of others, cannot be concealed. This naturally begets a general distrust of the whole, with that fatal lukewarmness warmnefs in Religion, which is its neceffary confequence.

The very habit of frequent reafoning, and difputing upon religious fubjects, diminifhes that reverence, with which the Mind would otherwife confider them. This feems particularly to be the cafe, when Men prefume to enter into a minute fcrutiny of the views, and œconomy of Providence, in the adminiftration of the world, why the Supreme Being made it as it is, the freedom of his actions, and many other fuch queftions, infinitely beyond our reach. The natural tendency of this is to leffen that awful veneration with which we ought always to contemplate the Divinity, but which can never be preferved, when Men canvas his ways with fuch unwarrantable freedom. Accordingly we find, amongft thofe fectaries where fuch difquifitions have principally prevailed, that he has been mentioned and even addreffed with the moft indecent and fhocking familiarity. The truly devotional fpirit, whofe chief foundation and characteriftic is genuine

nuine and profound humility, is not to be looked for among such persons.

Another bad effect of this speculative Theology has been to withdraw people's attention from its practical duties. — We usually find that those, who are most distinguished by their excessive zeal for opinions in Religion, shew great moderation and coolness as to its precepts; and their great severity in this respect, is commonly exerted against a few vices where the Heart is but little concerned, and to which their own dispositions preserved them from any temptations.

But the worst effects of speculative and controversial theology are those which it produces on the Temper and Affections. — When the Mind is kept constantly embarrassed in a perplext and thorny path, where it can find no steady light to shew the way, nor foundation to rest on, the Temper loses its native chearfulness, and contracts a gloom and severity, partly from the chagrin of disappointment, and partly from the social and kind Affections being extinguished for want of exercise. When this evil is exasperated by opposition and dispute, the consequences prove very fatal to the peace of society; especially when Men are persuaded, that their holding certain opinions entitles them to the divine favor; and that those, who differ from them, are devoted to eternal destruction. This persuasion breaks at once all the ties of society. The toleration of Men who hold erroneous opinions, is considered as conniving at their destroying not only themselves, but all others who come within the reach of their influence. This produces that cruel and implacable spirit, which has so often disgraced the cause of Religion, and dishonoured Humanity.

Yet the effects of religious controversy have sometimes proved beneficial to Mankind. That spirit of free enquiry, which incited the first Reformers to shake off the yoke of ecclesiastical tyranny, naturally begot just sentiments of civil liberty, especially when irritated by persecution. When such sentiments came to be united with that bold enthusiasm, that

severity

severity of temper and manners that diftinguifhed fome of the Reformed fects; they produced thofe refolute and inflexible Men, who alone were able to affert the caufe of liberty, in an age when the Chriftian world was enervated by luxury or fuperftition; and to fuch Men we owe that freedom, and happy conftitution, which we at prefent enjoy.—But thefe advantages of religious enthufiafm have been but accidental.

In general it would appear, that Religion, confidered as a fcience, in the manner it has been ufually treated, is but little beneficial to Mankind, neither tending to enlarge the Underftanding, fweeten the Temper, or mend the Heart. At the fame time the labours of ingenious Men, in explaining obfcure and difficult paffages of Sacred Writ, have been highly ufeful and neceffary. And though it is natural for Men to carry their fpeculations, on a fubject that fo nearly concerns their prefent and eternal happinefs, farther than Reafon extends, or than is clearly and exprefsly revealed; yet thefe can be followed by no bad confequences, if they are carried on with that modefty and reverence which the fubject requires. They become pernicious only when they are formed into fyftems, to which the fame credit and fubmiffion is required, as to holy writ itfelf.

We fhall now proceed to confider Religion as a rule of life and manners. In this refpect its influence is very extenfive and beneficial, even when disfigured by the wildeft fuperftition, as it is able to check and conquer thofe paffions, which reafon and philofophy are too weak to encounter. But it is much to be regretted, that the application of Religion to this end hath not been attended to with that care which the importance of the fubject required.— The fpeculative part of Religion feems generally to have engroffed the attention of Men of Genius. This has been the fate of all the ufeful and practical arts of life, and the application of Religion to the regulation of life and manners muft be confidered entirely as a practical art.— The caufes causes of this neglect seem to be these. Men of a philosophical Genius have an aversion to all application, where the active powers of their own Minds are not immediately employed. But in acquiring a practical art a philosopher is obliged to spend most of his time in employments where his Genius and Understanding have no exercise.— The fate of the practical parts of Medicine and of Religion have been pretty similar. The object of the one is to cure the diseases of the body; of the other, to cure the diseases of the Mind. The progress and degree of perfection of both these arts ought to be estimated by no other standard than their success in the cure of the diseases, to which they are severally applied.— In Medicine, the facts on which the art depends, are so numerous and complicated, so misrepresented by fraud, credulity, or a heated Imagination, that there has hardly ever been found a truly philosophical Genius, who has attempted the practical part of it. There are, indeed, many obstacles of different kinds, which concur

cur to render any improvement in the practice of phyfic a matter of the utmoft difficulty, at leaft while the profeffion refts on its prefent narrow foundation. Almoft all phyficians, who have been Men of ingenuity, have amufed themfelves in forming theories, which gave exercife to their invention, and at the fame time contributed to their reputation. Inftead of being at the trouble of making obfervations themfelves, they culled out of the promifcuous multitude already made, fuch as beft fuited their purpofe, and dreffed them up in the way their fyftem required. In confequence of this, the hiftory of Medicine does not fo much exhibit the hiftory of a progreffive art, as a hiftory of opinions, which prevailed perhaps for twenty or thirty years, and then funk into contempt and oblivion.— The cafe has been nearly fimilar in practical divinity. But this is attended with much greater difficulties, than the practical part of Medicine. In this laft, nothing is required, but affiduous and accurate Obfervation, and a good Underftanding

ing to direct the proper application of such Observation. But to cure the diseases of the Mind, there is required that intimate knowledge of the Human Heart, which must be drawn from life itself, and which books can never teach; of the various disguises, under which vice recommends herself to the Imagination; of the artful association of Ideas, which she forms there; and of the many nameless circumstances that soften the Heart and render it accessible. It is likewise necessary to have a knowledge of the arts of insinuation and persuasion, of the art of breaking false or unnatural associations of Ideas, or inducing counter associations, and opposing one passion to another; and after all this knowledge is acquired, the successful application of it to practice depends in a considerable degree on powers, which no extent of Understanding can confer.

Vice does not depend so much on a perversion of the Understanding, as of the Imagination and Passions, and on habits originally founded on these.

these. A vicious Man is generally sensible enough that his conduct is wrong; he knows that vice is contrary both to his duty and to his interest, and therefore all laboured reasoning to satisfy his Understanding of these truths is useless, because the disease does not lie in the Understanding. The evil is seated in the Heart. The Imagination and Passions are engaged on its side, and to them the cure must be applied. Here has been the general defect of writings and sermons, intended to reform Mankind. Many ingenious and sensible remarks are made on the several duties of Religion, and very judicious arguments are brought to enforce them. Such performances may be attended to with pleasure, by pious and well-disposed persons, who likewise may derive from thence useful instruction for their conduct in life. The wicked and profligate, if ever books of this sort fall in their way, very readily allow that what they contain are great and eternal truths, but they leave no lasting impression. If any thing can rouse them,

them, it is the power of lively and pathetic defcription, which traces and lays open their Hearts through all their windings and difguifes, makes them fee and confefs their own characters in all their deformity and horror, imprefles their Hearts, and interefts their Paffions by all the motives of love, gratitude, and fear, the profpect of rewards and punifhments, and whatever other motives Religion or Nature may dictate. But to do this effectually requires very different powers from thofe of the Underftanding. A lively and well-regulated Imagination is effentially requifite.

In public addrefles to an audience, the great end of reformation is moft effectually promoted, becaufe all the powers of voice and action, all the arts of eloquence may be brought to give their affiftance. But fome of thofe arts depend on gifts of Nature, and cannot be attained by any ftrength of Genius or Underftanding. Even where Nature has been liberal of thofe neceffary requifites, they muft be cultivated by much practice

before

before the proper exercife of them can be acquired.—Thus a public fpeaker may have a voice that is mufical and of great compafs, but it requires much time and labour to attain its juft modulation, and that variety of flexion and tone, which a pathetic difcourfe requires. The fame difficulty attends the acquifition of that propriety of action, that power over the expreffive features of the countenance, particularly of the eyes, fo neceffary to command the Hearts and Paffions of an audience.

It is ufually thought that a preacher, who feels what he is faying himfelf, will naturally fpeak with that tone of voice and expreffion in his countenance, that beft fuits the fubject, and which cannot fail to move his audience. Thus it is faid, a perfon under the influence of fear, anger, or forrow, looks and fpeaks in the manner naturally expreffive of thefe emotions. This is true in fome meafure; but it can never be fuppofed, that any preacher will be able to enter into his fubject with fuch such real warmth upon every occasion. Besides, every prudent Man will be afraid to abandon himself so entirely to any impression, as he must do to produce this effect. Most Men, when strongly affected by any passion or emotion, have some peculiarity in their appearance, which does not belong to the natural expression of such an emotion. If this be not properly corrected, a public speaker, who is really warmed and animated with his subject, may nevertheless make a very ridiculous and contemptible figure.— It is the business of Art to shew Nature in her most amiable and graceful forms, and not with those peculiarities in which she appears in particular instances; and it is this difficulty of properly representing Nature that renders the eloquence and action, both of the pulpit and the stage, acquisitions of such difficult attainment.

But besides those talents inherent in the preacher himself, an intimate knowledge of Nature will suggest the necessity of attending to certain external circumstances, which operate powerfully

powerfully on the Mind, and prepare it for receiving the defigned impreffions. Such in particular is the proper regulation of Church Mufic, and the folemnity and pomp of public worfhip. Independent of the effect that thefe particulars have on the Imagination, it might be expected that a juft Tafte, a fenfe of decency and propriety, would make them more attended to than we find they are. We acknowledge that they have been abufed, and have occafioned the groffeft fuperftition; but this univerfal propenfity to carry them to excefs, is the ftrongeft proof that the attachment to them is deeply rooted in Human Nature, and confequently, that it is the bufinefs of good fenfe to regulate, and not vainly to attempt to extinguifh it. Many religious fects in their infancy have fupported themfelves without any of thefe external affiftances; but when time has abated the fervor of their firft zeal, we always find that their public worfhip has been conducted with the moft remarkable coldnefs and inattention,

unlefs

unlefs fupported by well-regulated ceremonies. In fact it will be found, that thofe fects who at their commencement have been moft diftinguifhed for a religious enthufiafm that defpifed all forms, and the Genius of whofe tenets could not admit the ufe of any, have either been of fhort duration, or ended in infidelity.

The many difficulties that attend the practical art of making Religion influence the manners and lives of Mankind, by acquiring a command over the Imagination and Paffions, have made it too generally neglected, even by the moft eminent of the Clergy for learning and good fenfe. Thefe have rather chofen to confine themfelves to a tract, where they were fure to excel by the force of their own Genius, than to attempt a road where their fuccefs was doubtful, and where they might be outfhone by Men greatly their inferiors. It has therefore been principally cultivated by Men of lively Imaginations, poffeffed of fome natural advantages of voice and manner. But

as

as no art can ever become very beneficial to Mankind, unlefs it be under the direction of Genius and good fenfe, it has too often happened, that the art we are now fpeaking of has become fubfervient to the wildeft fanaticifm, fometimes to the gratification of vanity, and fometimes to ftill more unworthy purpofes.

The third view of Religion confiders it as engaging and interefting the affections, and comprehends the devotional or fentimental part of it. —The devotional fpirit is in fome meafure conftitutional, depending on livelinefs of Imagination and fenfibility of Heart, and, like thefe qualities, prevails more in warmer climates than it does in ours. What fhews its great dependence on the Imagination, is the remarkable attachment it has to Poetry and Mufic, which Shakefpeare calls the Food of Love, and which may with equal truth be called the Food of Devotion. Mufic enters into the future Paradife of the Devout of every fect and of every country. The Deity, viewed ed by the eye of cool Reaſon, may be ſaid with great propriety to dwell in light inacceſſible. The Mind ſtruck with the immenſity of his being, and with a ſenſe of its own littleneſs and unworthineſs, admires with that diſtant awe and veneration that almoſt excludes love. But viewed by a devout Imagination, he may become an object of the warmeſt affection, and even paſſion.— The philoſopher contemplates the Deity in all thoſe marks of wiſdom and benignity diffuſed through the various works of Nature. The devout Man confines his views rather to his own particular connection with the Deity, the many inſtances of his goodneſs he himſelf has experienced, and the many greater he ſtill hopes for. This eſtabliſhes a kind of intercourſe, which often intereſts the Heart and Paſſions in the deepeſt manner.

The devotional Taſte, like all other Taſtes, has had the hard fate to be condemned as a weakneſs, by all who are ſtrangers to its joys and its influence. Too much, and too frequent

frequent occaſion has been given to turn this ſubject into ridicule.—A heated and devout Imagination, when not under the direction of a very ſound Underſtanding, is apt to run very wild, and is at the ſame time impatient to publiſh all its follies to the world.—The feelings of a devout Heart ſhould be mentioned with great reſerve and delicacy, as they depend upon private experience, and certain circumſtances of Mind and ſituation, which the world can neither know nor judge of. But devotional writings executed with Judgment and Taſte, are not only highly uſeful, but to all who have a true ſenſe of Religion, peculiarly engaging.

The devotional ſpirit united to good ſenſe and a chearful temper, gives that ſteadineſs to virtue, which it always wants, when produced and ſupported by good natural diſpoſitions only. It corrects and humanizes thoſe conſtitutional vices, which it is not able entirely to ſubdue, and though it too often fails to render Men perfectly virtuous, it preſerves them

them from becoming utterly abandoned. It has befides the moft favorable influence on all the paffive virtues; it gives a foftnefs and fenfibility to the Heart, and a mildnefs and gentlenefs to the manners; but above all, it produces an univerfal charity and love to Mankind, however different in Station, Country, or Religion. There is a fublime yet tender melancholy, almoft the univerfal attendant on Genius, which is too apt to degenerate into gloom and difguft with the world. Devotion is admirably calculated to footh this difpofition, by infenfibly leading the Mind, while it feems to indulge it, to thofe profpects which calm every murmur of difcontent, and diffufe a chearfulnefs over the darkeft hours of Human Life.—Perfons in the pride of high health and fpirits, who are keen in the purfuits of pleafure, intereft, or ambition, have either no ideas on this fubject, or treat it as the enthufiafm of a weak Mind. But this really fhews great narrownefs of Underftanding; a very little reflection and acquaintance
with

with Nature might teach them, on how precarious a foundation their boafted independence on Religion is built; the thoufand namelefs accidents that may deftroy it; and that though for fome years they fhould efcape thefe, yet that time muft impair the greateft vigour of health and fpirits, and deprive them of all thofe objects for which at prefent, they think life only worth enjoying.—It fhould feem therefore very neceffary to fecure fome permanent object, fome real fupport to the Mind, to chear the foul when all others fhall have loft their influence.—The greateft inconvenience, indeed, that attends devotion, is its taking fuch a faft hold of the affections, as fometimes threatens the extinguifhing of every other active principle of the Mind. For when the devotional fpirit fall in with a melancholy temper, it is too apt to deprefs the Mind entirely, to fink it to the weakeft fuperftition, and to produce a total retirement and abftraction from the world, and all the duties of life.

I fhall

I shall now conclude thefe loofe obfervations on the advantages arifing to Mankind from thofe faculties, which diftinguifh them from the reft of the Animal world; advantages which do not feem correfpondent to what might be reafonably expected from a proper exertion of thefe faculties, particularly among the few who have the higheft intellectual abilities, and full leifure to improve them. The capital error feems to confift in fuch Mens' confining their attention chiefly to enquiries that are either of little importance, or the materials of which lie in their own Minds.—The bulk of Mankind are made to act, not to reafon, for which they have neither abilities nor leifure. They who poffefs that deep, clear, and comprehenfive Underftanding which conftitutes a truly philofophical Genius, feem born to an afcendency and empire over the Minds and affairs of Mankind, if they would but affume it. It cannot be expected, that they fhould poffefs all thofe powers and talents,

talents, which are requisite in the several useful and elegant arts of life, but it is they alone who are fitted to direct and regulate their application.

FINIS.

www.ingramcontent.com/pod-product-compliance
Lightning Source LLC
Chambersburg PA
CBHW032146230426
43672CB00011B/2465